# Designing Intelligence

# DESIGNING INTELLIGENCE

*A Framework for
Smart Systems*

STEVEN H. KIM

Massachusetts Institute of Technology

New York    Oxford
OXFORD UNIVERSITY PRESS
1990

Oxford University Press

Oxford   New York   Toronto
Delhi   Bombay   Calcutta   Madras   Karachi
Petaling Jaya   Singapore   Hong Kong   Tokyo
Nairobi   Dar es Salaam   Cape Town
Melbourne   Auckland

and associated companies in
Berlin   Ibadan

Library of Congress Cataloging-in-Publication Data
Kim, Steven Hyung.
Designing intelligence : a framework for
smart systems / Steven H. Kim.
p. cm.   Includes bibliographical references.
ISBN 0-19-5060161-4
1. Artificial intelligence.   I. Title.
Q355.K55   1990   006.3—dc20   89-48702

2 4 6 8 9 7 5 3 1

Printed in the United States of America
on acid-free paper

*To my parents*

# Preface

In recent decades, the study of intelligent systems has become increasingly vital and prevalent. Such systems have assumed a starring role in the analytic investigations of the scientist as well as the synthetic endeavors of the engineer and entrepreneur. Intelligent systems span a spectrum of human endeavors in disparate fields. They are found in niches such as the following:

- Adaptive structures of the biologist.
- Reasoning mechanisms of the cognitive scientist.
- Microprocessor-based devices of the engineer.
- Automated factories of the industrialist.
- Management information systems of the executive.
- Global communication networks of the sociologist.

Despite the growing importance of intelligent systems, we lack a unifying theme: there is not yet a general theory nor even a coherent framework for discussion.

This book presents a partial remedy by proposing a framework for intelligent systems. The framework provides a checklist of parameters that may be invoked for both analysis—the study of the nature and limitations of systems—and synthesis—the construction of intelligent systems.

The rise of the scientific method, followed by the rapid pace of technological advances, led to the Industrial Revolution and the Information Society. During that period, the mechanistic view has prevailed and held sway over designed systems. In fact, both theoretical and empirical endeavors in the realms of science, engineering, and design have become largely synonymous with the mechanistic view.

However, there is a growing realization that the organic or biological perspective has much to offer, not only as a domain of study, but also as a metaphor for artificial objects. For example, engineered systems should be more malleable than brittle, more forgiving than demanding, more holistic than fractional. Artificial objects, like natural things, should move gracefully rather than clumsily, tolerate imprecision rather than require perfection, fail softly rather than catastrophically, repair themselves rather than decay, and improve over time rather than stagnate.

For this reason, a fair share of the examples in this book are drawn from the biological sphere. The general framework to be described, however, is intended to serve in the investigation and construction of both artificial and natural systems.

This book originated in part from lecture notes prepared as introductory material for several courses offered at the Massachusetts Institute of Technology in 1987 and 1988. These courses, spiced with an interdisciplinary flavor, ranged from an undergraduate seminar (Introduction to Intelligent Systems) to a graduate subject (Topics in Knowledge Engineering). The framework described in this book has been used as the conceptual foundation for design projects in a spectrum of topics: from mobile robots to sentient houses, from intelligent prosthetics to factory supervisors, and from poem synthesizers to knowledge-based marketing tools. In addition to the introductory notes, each course was fortified by more advanced readings appropriate to the subject at hand. I am indebted to the students in those courses for stimulating discussions.

A number of other people played vital roles during the preparation of the manuscript. Yvonne Cheung assisted in the preparation of the biological examples for the initial set of class notes. Elias Towe and K. Eric Drexler reviewed earlier drafts of the manuscript. Margaret Herbig, Mary Jane Close, Lindsay Moran, and Kristen Svingen provided able assistance with the proofreading and text editing. Their efforts were enhanced by the contributions of Elizabeth Ellis and Polly De-Frank. The exhibits in the book represent a joint effort of several people, including Lindsay Moran and Maureen Kelly; Figure 11.7, in particular, was drawn by Christopher Wilson.

To these individuals I am grateful for the comments and suggestions for improvement, both large and small. Any remaining errors and feebleness of argument are, of course, my own handiwork.

*Cambridge, Mass.*                                                                          S.K.
*January 1990*

# CONTENTS

# I

# PRELUDE

*If, therefore, anyone wishes to search out the truth of things in serious earnest, he ought not to select one special science; for all the sciences are conjoined with each other and interdependent.*[1]

René Descartes

# 1

# Introduction

*Natural philosophy consists in discovering the frame and operations of nature, and reducing them, as far as may be, to general rules or laws—establishing these rules by observations and experiments, and thence deducing the causes and effects of things.*[1]

Isaac Newton

The phenomenon of intelligence seems easier to recognize than to define. This elusive concept has attracted the attention of many a student in fields ranging from the physical sciences to the social and life sciences. Investigators in biology, psychology, and linguistics have attempted to analyze it; students in computer science, engineering, and management have tried to emulate it. Philosophers have debated whether it can be fathomed at all, let alone fabricated.

In the midst of this nebulosity, however, one thing seems clear: intelligence is taking a more prominent place in our collective consciousness. This trend has resulted primarily from advances in the biological sciences, artificial intelligence, and computing technology.

Most important, perhaps, we have acquired in recent decades the capability to fulfill the age-old human aspiration to create intelligent entities. To an increasing degree, the operant question is not "Can we create intelligence?," but rather "How intelligent?"

A science of intelligent systems should consist of a series of layers for describing intelligent phenomena. These levels reflect a progression from the conceptual and qualitative to the rigorous and quantitative. The knowledge base consists of the following layers:

- A *framework* for discussion, including a systematic set of concepts, an identification of critical issues, and a consistent terminology.
- A *model* describing the relationships among the objects of the framework. The model provides a rigorous representation of the objects and their relationships, and at the same time accommodates techniques for manipulating or utilizing the formal representations. The general model may consist of a number of submodels for describing limited classes of intelligent phenomena.
- A *theory* consisting of a set of principles describing the fundamental behavior of such systems, as well as implications for their synthesis. The theory should be expressed in terms of the general framework and its referent models.

3

To illustrate, a conceptual framework for mobile robots consists of a set of terms relating to its salient features, such as degrees of motion and modes of sensory input. The corresponding formal framework will involve a set of abstract symbols and their association with the basic concepts of the informal description.

In contrast, a model builds on the framework and allows for the description of the relationships among the basic objects. An example of this lies in the modeling of the impact of sensory deprivation on mobility or overall system performance.

Finally, a theory organizes a model by affixing a set of principles relating to the operation or behavior of the referent system. For example, a theory should be able to deduce the consequences of a robot's behavior when the system is faced with alternative choices or even conflicting objectives. By constraining a model with the inferences or relationships which are permissible, a theory offers the advantages of predictive power and behavioral description.

This book addresses these three imperatives for a science of intelligence. It presents a general framework for the analysis and synthesis of adaptive systems, as well as a set of specific principles and design guidelines. The framework is a multifaceted structure that is outlined in the introductory sections to follow, and described in greater depth in Parts II and III along with a number of design principles.

The appendices present a set of models and mathematical tools that may serve as the kernel for a comprehensive theory of intelligent systems. The full story, however, remains to be developed and must wait its turn to be told at some future date.


## Overall Framework

A framework for an interdisciplinary study of intelligent systems should be comprehensive, versatile, and minimal. In other words, the framework should accommodate all the relevant topics, allow for concepts at varying levels of detail, and be nonredundant.[2] The overall framework for the analysis and synthesis of intelligent systems is given in Figure 1.1. This framework shows four directions: *arena, phase, aspect,* and *factor.*[3] Each of these directions may be viewed as an independent dimension along which to engage in the study of intelligent systems. Each axis of the four-dimensional framework may in turn be partitioned into lower-level issues.

The *natural* arena refers to systems occurring freely in nature, without human intervention. In contrast, the *artificial* arena relates to engineered objects. With continuing advances in molecular biology and intelligent prosthetics, the distinction between natural and artificial functions will become increasingly blurred.

The *analysis* phase refers to the development of tools and theories for studying the nature and limits of intelligent systems, while the *synthesis* phase relates to prescriptive decision rules for designing better devices. An example of an analysis issue is the study of information characteristics and their impact on system performance, or the investigation of logic as a unifying theme for reasoning processes. An example of a synthesis issue is a set of principles for matching internal architecture to system objectives.

The third category relates to *theoretical* and *practical* aspects. The *theoretical*

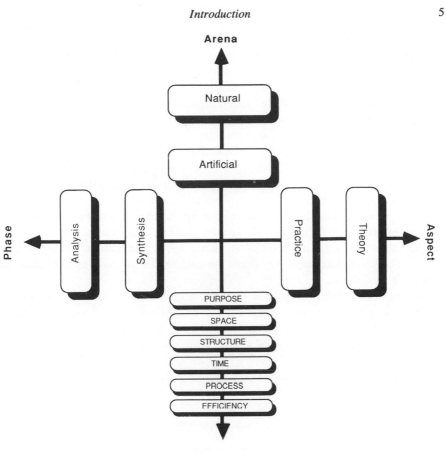

**Figure 1.1** Overall framework for intelligent systems.

aspect pertains to general scientific principles that hold across the entire spectrum of intelligent systems, while the *practical* aspect relates to tools, heuristics and decision support systems applied to the construction, operation, or maintenance of specific devices. An example of a practical issue at the micro level is the efficient integration of visual and tactile data for robotic motion. A macro issue, on the other hand, relates to design and implementation strategies for realizing multi-agent systems.

The fourth direction relates to the *factors* of intelligence. These are independent characteristics that define the nature of an intelligent system. The parameters of intelligence—*purpose*, *space*, *structure*, *time*, *process*, and *efficiency*—will be the subject of detailed discussion in later chapters.

These four dimensions may be isolated for the sake of discussion, but they are often interdependent. For example, it would make little sense to investigate the requirements for the hardware component without taking into consideration software concerns such as behavioral specifications.

Since the overall framework is a multidimensional structure, it may be decomposed into any convenient subframework along one or more axes. This book focuses primarily on the topic of attributes, which are further defined and explored in Parts II and III. In other words, the theoretical and practical implications of the attributes for intelligence are discussed in terms of the analysis and synthesis of systems in both their natural and artificial forms. An equally valid approach would be to elaborate on the framework along any of the other axes relating to phase, arena, or aspect.

## Systems

If we are to discuss intelligent systems, perhaps some preliminary remarks about systems are in order. The physical scientist views a system as any subset of the material universe. Such a person may place a conceptual box or *control volume* around a set of billiard balls or within a flowing stream of fluid and designate the contents as a system.

On the other hand, the general systems scientist defines a system in a more abstract way. To the systems specialist, a system is any collection of interrelated objects, whether physical or abstract. According to this perspective, a set of robots may constitute a system just as well as a collection of equations or even the set {life, liberty, happiness}.

Our concern in this book is with systems that have some ultimate physical embodiment. Hence the physical scientist's view of a system is more fitting for much of the discussion. For this reason, we define a *system* as some collection of interrelated objects, as contrasted with the *environment* which refers to the rest of the universe.

A system consists of parts that interact with one another. Through these interactions and their relationships to the environment, the components may give rise to *emergence*, or the genesis of systemic behaviors that transcend those of the individual constituents. Emergence embodies the concept "The whole is greater than the sum of its parts." Specifically, an *emergent* property of a system is one that cannot be ascribed to any of its individual components.[4] An example is found in the wave phenomena that occur within a large collection of gas particles. Sound waves, best regarded as continuous phenomena having specific directions of travel, are propagated through a medium consisting of discrete particles that individually travel in all directions.

Another example of emergence is instability arising from stable components; although each part may be steady by itself, the system as a whole may exhibit precarious tendencies. This happens when the controller for an autonomous vehicle overcompensates for an error in one direction, leading to a wild swing to the opposite side, then back again and so on, ultimately resulting in loss of control and system failure. In fact, the suppression of instability is a central concern in the field of control systems engineering.

In general, a system as a whole possesses characteristics that have little meaning when attributed to its individual components. This is especially true of nonlinear characteristics. For example, an automotive factory might be distinguished by a unit production rate of 50 cars per robot per day, but a per capita index of the quality of cars produced by the factory would be less meaningful as a descriptive characteristic.

## Purpose, Intelligence, and Learning

Although intelligence is a central concern at institutions of research around the world, it has so far eluded a widely-accepted definition. One dictionary defines intelligence as "the ability to understand or to deal with new or trying situations." However, imprecise terms such as "understand" or "deal with" leave much to be desired. Can a weather vane be considered intelligent when it operates in a sandstorm, a "new and trying situation" beyond its previous range of exposure?

Perhaps it is simpler to determine what intelligence is, in terms of what it is not. Intelligence is not a physical object, but a concept or an abstraction; not a device or a configuration, but rather a process or a behavior.

Hence, a system embodies intelligence by virtue of the fact and to the extent that it displays intelligent behavior. The physical embodiment or the organizational structure of a system may be crucial to the support of intelligent behavior, but neither factor of itself can be said to be intelligent.

Intelligent behavior manifests itself through some physical objects. These objects are composed of atoms and molecules to which we do not ascribe intelligence. However, they serve as building blocks for objects that can exhibit remarkable levels of sophistication. Intelligence is a matter of degree, spanning the spectrum from the null to the highly intelligent. In fact, even intelligent systems are composed of unintelligent parts. This viewpoint is embodied in the work of Marvin Minsky, who has explained the workings of the human mind in terms of a community of agents.[5] Like ants in a colony, each agent has limited capabilities, but through interaction with fellow agents, achieves collective results that transcend individual efforts.

The concept of purpose seems to be related to intelligence; we would be reluctant to call an object intelligent if it exhibited purely random and unintentional behavior. An example is a particle of smoke drifting in the wind. One apparent exception to the randomness principle fortifies the rule: software for generating a list of "random" numbers has the purpose of producing a numeric list that may then be used to model processes having certain definite and recognizable characteristics.

We can readily attribute some specific goal or purpose to engineered systems, since they are generally constructed to fulfill a need. Two examples of this are the thermostat whose purpose is to regulate room temperature, and the automatic pilot that flies a plane.

In addition, the behavior of natural systems can be understood more readily if they are associated with some goal. To wit: the purpose of an animal's neural system is to transmit information, while its survival instinct ensures its physical preservation. The foregoing discussion is encapsulated in the following proposition:

- A *purposive* system is one that serves a discernible goal, function, or objective.

In this vein, a wire sculpture of arbitrary shape may be regarded as a device whose purpose is to express an artist's creative thought.

We would hesitate to call a system intelligent if it were only capable of behaving in simple ways bound by the limitations of a sparse set of internal states. An intelligent agent must be capable of displaying a rich set of responses.

- An *intelligent* system is one that exhibits a rich behavior space in the pursuit of its objectives.

The class of intelligent systems is therefore a subset of purposive systems.

The richness of the behavior space implies that the system is capable of a wide repertoire of behaviors relating to its goals. The nature and extent of the behavior space determine the degree of intelligence.

The behaviors define a partial ordering among intelligent systems, so that one entity may be considered to be of greater, equal, or lower intelligence than another. According to this view, a numeric calculator with the four functions of addition, subtraction, multiplication, and division is more intelligent than another featuring fewer functions.

A purposive system that evolves over time is called a learning system. Learning enables an agent to modify its repertoire of behaviors to respond more effectively to environmental conditions.

- A *learning* system is a purposive entity whose behavior space changes over time to better fulfill its objectives.

A learning system, by definition, will respond differently to similar stimuli over some stretch of time.

The learning phenomenon may be as sophisticated as the induction of generalized heuristics from specific instances; this is illustrated by the ability to distinguish poisonous and nutritious herbs. The learning may be simple, as in the accommodation of breathing to a rarified atmosphere or gradually growing used to a distracting noise.

Intelligence and learning are correlated, but not coincident, attributes. For example, a conventional information system for managerial control may be considered intelligent; but its behavior specification is fixed and therefore not learning. On the other hand, the development of callouses in response to walking barefoot might be considered a learning phenomenon at the physiological level, since it enhances an organism's ability to negotiate the terrain; but it would not be regarded as an intelligent one. The relationships among purposive, intelligent, and learning systems are depicted in Figure 1.2.

A rich external behavior space must be supported by an equally diverse set of internal states. This correspondence is explored further in Chapter 4 on the relationship between a system and its environment.

We may call a system's behavior *adaptive* if it serves to attain a goal in the face of changes in the environment or the system itself. That is:

- An *adaptive* system is one that is capable of changing its behavior to attain a goal, despite environmental changes that would otherwise thwart its purpose.

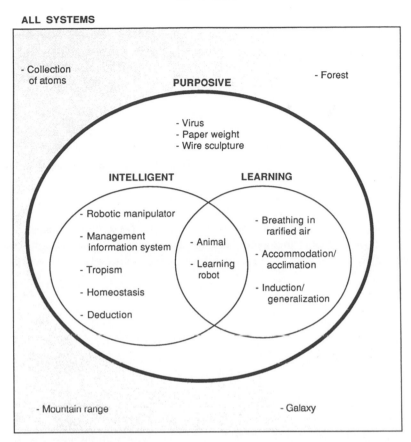

**Figure 1.2** Relationships among purposive, intelligent, and learning systems.

A heliotropic plant that orients itself to the sun's trajectory across the sky is an adaptive and purposive system. On the other hand, a paperweight would be considered a purposive but not an adaptive system.

The term *adaptive* has been used by previous writers to refer to either intelligence or learning. The first interpretation holds when a thermostat or control system is called adaptive; it is intelligent in the sense that the system reacts to a variety of external stimuli, but its response to these inputs remains fixed over time. The system is therefore intelligent but not learning.

According to the second interpretation, adaptation refers to learning. When an animal enters a new forest after depleting sources of food in its previous territory, it must adapt to the new environment. A leopard that ruled a jungle devoid of lions cannot afford to remain oblivious if its new territory is populated by them. In this context, adaptation refers to learning.

In a similar way, an organization must react to changing environmental circumstances. A monopolistic vendor cannot maintain high prices indefinitely if it is subject to reprisal from regulatory agencies or challenge from potential competitors.

A consumer electronics company cannot afford to rely on a five-year product development cycle when competitors introduce new designs in pace with microprocessor technology, where three years separate one generation from the next.

The learning interpretation will be used for *adaptation* throughout this book. In other words, an adaptive system is a learning entity; it may or may not be intelligent.

Many sophisticated systems, such as animals and smart robots, are both learning and intelligent. Often adaptive mechanisms are required to support intelligent activities. For example, body temperature in a warm-blooded animal must be carefully regulated to maintain physiological processes, without which the animal would be dead and not very intelligent.

Our collective understanding of adaptation and intelligence is yet to be crystallized. The "definitions" given above for adaptive and intelligent entities should be regarded as working definitions for the scope of this book, rather than generally accepted dogma. Hence, intelligent behavior is a matter of degree, ranging from a simple arithmetic program to the human brain, which epitomizes the concept.

## Organism and Purpose

Perhaps the attribution of goals to animate objects deserves further discussion. It is not clear whose Design and Purpose are served by natural organisms. However, whatever their "real" role, we may attribute to such goals an operative role that will suffice for the purposes of the current discussion. The molecular biologist Salvador Edward Luria has written of the lack of purpose in evolution. For example, a particular visual system or mechanism for gene regulation is a workable design that is adequate but not necessarily optimal for its role. In fact, all biological structures are the progeny of a "purposeless historical process."[6]

But why is a particular "exceptional sample" selected over others? The Darwinian notion of "survival of the fittest" stipulates a culling mechanism for choosing one characteristic or even species over another. In some sense, we may attribute a purpose of survival to biological organisms: survival of the self and of the species.

To elaborate on this idea, we may draw on a concept from management science. The "satisficing" theory of human behavior states that decisions are based on choices that yield a satisfactory measure of utility. Due to limitations in human memory and computation, a decision maker can, in general, perform neither a complete enumeration nor evaluation of all the potential choices.[7] By fusing these concepts, we may regard the hand of evolution as a "satisficer," one whose effective purpose is to select a mechanism to ensure the survival of a species in the face of changing environmental conditions. In this way, the hand of evolution picks out a satisfactory but not necessarily optimal set of characteristics.[8]

To date, the course of evolution has relied largely on the probabilistic process of genetic mutations. Once a course is set, it is difficult to switch to a completely new path or even to revert to an earlier, simpler state.[9] For instance, a prototype vision system that responds feebly to light will be developed further by the same environmental forces that engendered the prototype in the first place. Unless the environment changes drastically, selective survival will ensure that the species will not revert to its earlier state and lose the capacity for light detection. In addition, the species will not likely develop a visual system based on fundamentally different principles, for that would require too many mutations of the proper character occurring all at once. In this way, natural evolution acts as an incremental optimizer, enhancing biological structures one step at a time in response to environmental pressures, without the benefit of a global plan nor of the knowledge of the entire universe of possibilities.

The English zoologist Richard Dawkins has advanced the view that an animal is a gene's vehicle for replicating itself. Since many genes have survived in a competitive and sometimes ruthless environment, it is only natural for them to be self-serving, at the expense of other genes if necessary.[10] In this way, even genes may be attributed with an innate purpose.

Whatever may be the objectives of the evolutionary process, individual animals generally behave as if driven by goals. Foremost among these is the goal to survive, followed by objectives of lower priority such as shelter and comfort.

## Information Processing for Intelligence

As is apparent from our previous definition, the difficulty of capturing the nature of intelligence derives from its hydra-headed quality. Since the environment can change in myriad ways, an intelligent system should be capable of a like number of adaptive responses.

If a system is to respond to a dynamic environment, it must be able to detect external changes. Recognized changes constitute information that may be utilized immediately by the system or recorded for subsequent use.

The detection of external information and its subsequent recording are separable concepts; sometimes the external input leaves a lasting effect on a system and at other times not. This is due in part to the physical properties of materials. For example, a piece of bone that undergoes compression will regain its previous shape if the original deformation did not exceed the bone's elastic limit; if this limit is exceeded, the bone will exhibit a lasting record of the compression. A reflex action in response to a sudden noise exemplifies the immediate use of external information. In addition, data concerning the noise may be stored for the guidance of future action.

The storage and processing of information are hallmarks of intelligent systems. Often, we think of information storage as *memory* when it occurs in animate objects or computer systems.

To the extent that the accrual of information manifests itself in a detectable physical change, we may view information as being processed at all times. A question that often comes to mind is whether or not a system processes information. In general, however, the relevant question is not, *"Is* it processed?" but, "By *how much*?" For example, the mechanical vibrations in sound are transformed into electric signals in the human brain. In a similar way, the profusion of light impulses impinging on the retina is first aggregated there, then passed on to the visual cortex as voltage signals.

Viewed in these general terms, the storing and processing of information are pervasive in both nature and artifact. These phenomena support complex information processing functions.

In high-level systems such as primates, adaptive behavior is supported by a portfolio of capabilities ranging from reasoning to learning and communication. To return to an earlier line of thought: reasoning may be viewed as an advanced form of processing; learning may be regarded as the functional harmony of memory and reasoning; communications may be considered as the coordination of memory and reasoning among individual systems.

In summary, we may loosely define intelligence as a set of behaviors that facilitate adaptation and survival within a disinterested environment. These behaviors relate to the perception, modeling, and manipulation of the system and its accompanying surroundings. Such activities are supported by memory, reasoning, learning, and other deliberate capabilities needed to perform effectively not only in the present, but in anticipation of the future.


## Varieties of Intelligence

Figure 1.3 shows a classification of intelligent systems. In the natural realm, the plant kingdom offers few examples of behavior that we would call intelligent. An exception might be the feeding behavior of insectivorous plants such as the Venus flytrap. Another adaptive example occurs when a sapling branch that is subjected to a moderate, continuous load breaks only at a much greater weight compared to another that has had no prior conditioning. This type of behavior may be viewed as the product of intelligent behavior in slow motion. There is little learning within the life span of an individual plant, but minute changes take place over evolutionary time scales that result in discernible behavior.

But our interest is centered on more impressive repertoires of intelligent action. Therefore, most of our examples in the natural arena will be culled from the animal kingdom, and will tend to focus on human subsystems.

The artificial realm of intelligence may be conveniently partitioned into planned and unplanned systems. The former class pertains to objects of deliberate human design; these in turn may be further divided into the categories of software, hardware, and organizational structures. Although these groups serve as suitable categories for discussion, they are not mutually exclusive. For example, a large-scale organization such as a multinational corporation contains both hardware and software structures (including people, who embody both!).

The class of unplanned systems denotes objects that result from human action but are not consciously designed. An example in this category is the class of natural languages; another is the institution of reciprocal commitments to sustain a community in a feudal society.

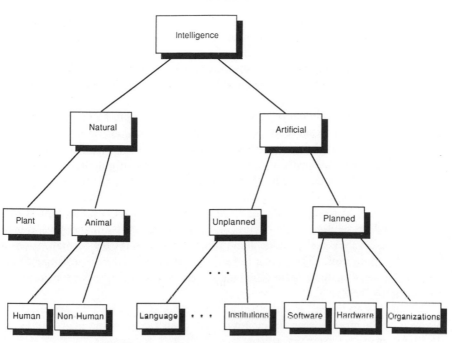

**Figure 1.3** Typology of intelligent systems.

Unplanned systems, almost by definition, are fashioned to a large degree by natural forces. Hence the distinction between natural and artificial systems may be better viewed as extreme points on a continuous scale rather than nominal categories.

## Scope of Book

This book presents a general framework for analyzing and synthesizing intelligent systems of both natural and artificial persuasion. This introductory chapter has explored the nature of intelligence and its manifestation in a variety of forms. The study of intelligent systems may be partitioned into three stages: framework, model, and theory. To this end, a general framework is presented and interpreted in the context of various applications. A number of rigorous models are also discussed in the Appendices. Further, a preliminary set of principles are proposed that may serve as a basis for a theory of intelligent systems.

The second chapter in Part I presents an overview of the general framework for intelligent systems, drawing on natural as well as artificial systems. The nature of the dimensions as well as their mutual relationships are briefly discussed.

Part II presents, in greater detail, each factor of the framework. These attributes range from space and time to process and efficiency, and may in turn exhibit important subissues, as exemplified by the trade-off between flexibility and tractability or the interplay between short-term and long-term pursuits. Each chapter discusses the relevant *intra*factor trade-offs.

Part III explores a number of *inter*factor trade-offs and their relationships to some commonly used concepts. Among these dichotomies are mechanism versus process, and time versus space.

Part IV presents a number of case studies. The first is a product of human scale, an autonomous robot to serve as a general-purpose helper. The second is a larger engineering project involving the construction of an advanced manufacturing plant. The factory is to be efficient and capable of speedy reconfiguration to adapt to dynamic market demands. The last application deals with the redesign of an organization in the face of a crisis precipitated by environmental challenges.

Finally, Part V wraps up the discussion and is followed by the appendices. These appendices explore some interesting issues in greater detail and have a more technical flavor; they also provide some future directions for the formalization of the general framework, extensions toward a systematic theory of intelligent systems, and a set of mathematical tools which may be instrumental for the construction of such a theory.

An initial step toward a formal model of design is presented in Appendix A, followed by the closed nature of creativity and deliberate thinking in Appendix B. The third appendix discusses a general methodology for design based on the requirements of functional independence and information minimization. Appendix D presents a formal framework for learning systems, and Appendix E, a generalized measure of information based on the probability of attaining a set of objectives. These appendices highlight a number of promising directions for future work.

The last appendix illustrates the utility of the framework at varying levels of description. It involves the case study of a knowledge-based system to design flexible manufacturing plants.

A theory of intelligent structures must fuse the frameworks and models with a set of relationships or behavioral principles. Some ideas along these lines are discussed in Part II in connection with the exploration of individual dimensions. For example, Chapter 7 examines the proposition that the effective level of centralization depends in part on the speed of response required of a system; more specifically, the appropriate degree of control rises with the response rate, then eventually declines.

The design principles discussed throughout the book are relatively domain-independent observations that apply to a wide spectrum of intelligent agents. For any specific area of application, these general principles will be augmented by domain-dependent knowledge, whether in the form of energy storage devices for mobile robots, or environmental factors for a marketing organization. A number of theoretical and practical considerations for the fusion of domain-dependent and domain-independent knowledge are discussed in Appendices A and B.

The frameworks, principles, and theories for intelligence will evolve in the years to come. The refinements and extensions will be the result of active investigation into the general phenomenon of intelligence as well as our collective experience in constructing intelligent entities of increasing sophistication.

# 2

# Overview of Framework

*From the war of nature, from famine and death, the most exalted object which we are capable of conceiving, namely, the production of the higher animals, directly follows. There is grandeur in this view of life. . . . from so simple a beginning, endless forms most beautiful and most wonderful, have been and are being evolved.*[1]

Charles Robert Darwin

The design of intelligent systems requires the consideration of a number of factors or dimensions. This chapter first presents an overview of these factors, including a discussion of interdimensional trade-offs. A discussion of the relationships among the design attributes, including their relative independence as well as the trade-offs among pairs of factors, concludes the chapter.

## Factors of Intelligent Systems

Intelligent systems exhibit a number of common features such as the need to coordinate activities among multiple elements and to implement them on physical structures. These characteristics fall into half a dozen clusters or dimensions relating to purpose, space, structure, time, process, and efficiency. A schematic of these dimensions is given in Figure 2.1. The dimension of purpose refers to the goals of an intelligent system and relies on the other five factors for its fulfillment.

### Purpose

Chapter 1 has presented the viewpoint that intelligence is defined by behavior rather than by a structure or physical entity. Moreover, we would hesitate to ascribe intelligence to purely random or purposeless behavior. According to this perspective, purpose is a hallmark of intelligence.

The dimension of purpose may be characterized by the effectiveness with which a system's objectives are met. In a shifting environment, the goals of a system may change over time; in this context the ability to fulfill goals will be enhanced by the versatility of behaviors available to the system.

In the pursuit of its goals, a system must possess some knowledge of procedures

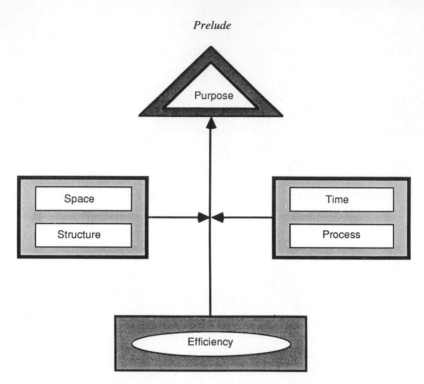

**Figure 2.1** Factors of intelligent systems. The purpose of a system is supported by the five other factors.

as well as information about the environment. Such knowledge may be represented explicitly as in the form of decision rules, or implicitly as in the hardware of a mechanical calculator. Which particular processes are activated within the system depends partly on the environment, whose status is represented by the system in the form of data and information.

## Space

Intelligent behavior manifests itself in some finite region of space; usually this region is simple to define, but at other times the precise boundaries may be subject to debate. These perimeters, once prescribed, define the extent of the system on the interior and the expanse of the environment on the exterior. By appropriately defining the system and its interface to the environment, it is possible to implement intelligent behavior as a shared phenomenon between the system and its environs.

## Structure

Our concern in this book lies in intelligent systems that take some form of physical expression. Such objects are characterized by some pattern or structure. For example, a monolithic slab of homogeneous material is unlikely to exhibit intelligent behavior. Hence it seems reasonable to assert that any system can be partitioned into identifiable components or modules. The basic units of the system then define its coarseness or granularity.

Sometimes it is convenient to think of granularity in terms of continuity, whether of a structure or a process. For example, a system may monitor an environmental condition at discrete time intervals or on a continuous basis.

For the granules of the system to interact with each other in coherent fashion, they must take some form of organization. The network is an organizational structure consisting of a set of interrelationships among arbitrary pairs of modules. A specialization of the network structure is the hierarchy, in which system components are grouped into strata or levels. Each module may interact with any number of elements in the level below, but with no more than one in the level above.

The layer architecture resembles the hierarchy in its clustering of components by strata. A granule in a layered structure, however, communicates through standard interfaces or protocols to the levels both above and below its own. Hence this architecture eliminates the need to define one-to-one relationships among modules in separate strata.

Another feature of system structure relates to parallelism, the degree to which separate processes occur simultaneously. Biological systems often incorporate parallel processes, as in the human visual apparatus, where incoming light is sensed through numerous photodetectors, processed concurrently, then conveyed to the visual cortex.

To date, however, engineered devices have usually taken refuge in the simplicity of a single stream of processes. But the limitations of single streams for expressing a rich set of behaviors in reasonable time suggest the need for parallel architectures. An example of parallelism is found in computer structures which incorporate thousands of processors acting in unison.

*Time*

Intelligent behavior transpires over a certain period of time. Since the environment is often in continual flux, the system must respond promptly to accommodate the changes.

Although many activities are dynamic, some aspects of them are static or unchanging over time. This is true, for example, of regulatory processes. The task of a chemical refinery controller, for example, is to maintain a steady level of production.

In addition to environmental factors, a system must combat its own tendency toward entropy or decay. These effects may be due to steady factors such as wear and tear, or unexpected occurrences such as voltage surges or mechanical collisions.

A special consideration relating to timeliness is the dichotomy between realtime and batch, also referred to as online versus offline processing. Realtime activities refer to the formulation of tactics and the implementation of decisions in response to conditions as they arise; batch refers to the acquisition of data for processing at a future date. For some activities, the choice of realtime versus batch is clear-cut. The shutdown procedures for a nuclear reactor malfunction should operate in realtime;

but the data analysis for astronomical observations may be performed in batch mode.

As with many other attributes, the distinction between realtime and batch is a matter of degree. To the extent that computational processes take time, both in biological and artificial systems, no decision can be made instantaneously and must therefore be a batch activity.

The dimension of time also introduces the dichotomy between the now and the later. Under the constraint of finite resources, a trade-off must often be made between performance in the present versus the future. For example, a machine might be operated at maximum capacity today at risk of failure tomorrow. Capital may be spent in a similar way to enhance the corporate income statement this year, or invested for the sake of the future balance sheet.

In a universe of predictability, a system may be designed rationally from the beginning and expected to behave as foreseen. That universe is not ours, however, for all our days seem indeterminate.

The lack of predictability in events may result from some probabilistic features inherent in nature.[2] Or it may arise, as Einstein believed, from our incomplete understanding of the world.

Whatever the cause, the operational consequence is the same. In particular, the system architect must make decisions based on the assumption that the object of his design, as well as its operating milieu, is nondeterministic. An effective design, then, should be able to accommodate likely sources of stochastic phenomena. These may originate from within the system, as in sensor failures, or outside it, as in the appearance of unexpected obstacles.

## Process

An intelligent process consists of a sequence of actions or events. These activities may be characterized by the level of initiative. At one extreme, a system may initiate all interactions with its environment; at the other, the system may wait passively and react only in response to activities generated externally. In fact, a sentinel system might merely observe, performing no direct action upon its environment except to consume energy and thereby release waste heat products, or eventually fall into disrepair.

A process may also be characterized by its direction of progression. It may move forward from its initial state, exploring promising avenues of inquiry; or it may move in the opposite direction, from its end point to the initial state, in search of its goal.

Another aspect of directionality is that of enhancement versus reduction, or construction versus consolidation. On one hand, a system may begin with a modicum of knowledge and construct a full-fledged model or strategy based on this alone. The other approach to enlightenment involves the consolidation of existing structures into simpler structures: an example is found in the reduction of large amounts of data into summary statistics such as the mean and variance.

For a set of processes to achieve some useful end, it must be coordinated in conjunction with the physical mechanisms that serve as its instruments of expression. An important facet of such coordination is feedback, consisting of a closed-loop or open-loop scheme. In a closed-loop approach, the output or result of a process is monitored to determine whether the system objectives are met. In an open-loop scheme, a set of processes are activated and trusted to yield the desired results.

The closed-loop approach obviously involves more complexity than the alternative. The added complexity and cost may or may not be justified by the improvement in performance. Many systems of reasonable complexity will likely incorporate subsystems of both types.

Another feature of coordination relates to centralization, the degree to which processes are orchestrated from a central point or distributed over time and space. Centralized control tends to be effective in a dynamic situation where changes occur rapidly, and must be responded to with equal speed. This is often true in the tactical environment of a battlefield or in the commercial markets of products which incorporate rapidly-advancing technologies.

On the other hand, the rate of change may be too swift, or the amount of information too voluminous to be dealt with by a centralized response. The commander-in-chief may dictate general strategy, but is incapable of generating a comprehensive response to battlefield conditions in all theaters of war.

In a similar manner, the chief executive of a major corporation can only develop policy; the details of its implementation must be relegated to the various divisions. In the juxtaposition of centralized and decentralized functions, biological systems present numerous examples. To illustrate, the brain procures information and serves as the center of reasoning and planning; but the reflex action needed to withdraw a finger from a flame or a toe from a thornbush is relegated to distributed centers of processing in the form of neural chains centered in the vertebrae.

## *Efficiency*

A design may be judged perfect by each of the preceding dimensions. However, it will still be only of academic interest if it is impractical to implement. This impracticality may result from an excessive need for resources, whether in initial fabrication or in the requirements of operation or maintenance.

Such costs may be explicit, as in the monetary cost of constructing a vehicle; or they may be implicit, as in the inefficient use of energy conversion in a solar panel or a biomedical device. For these reasons, efficiency often determines the feasibility of a design and its ultimate utility.

## Relationships Among the Factors

The previous section has presented the design dimensions for intelligent systems, as well as some intradimensional trade-offs. The foregoing discussion illustrates some of the concerns that must be resolved, whether explicitly or implicitly, in the design of an intelligent system.

The attributes of the general framework, along with examples in each class from

*Prelude*

both the natural and artificial realms, are listed in Tables 2.1 and 2.2. The design attributes and the examples will be discussed in greater detail in Part II.

## Independence

The attributes for intelligent systems are conceptually separable for convenience in discussion and system evaluation. In general, they are independent in the sense that one attribute does not fully cover the issues incorporated in another.[3]

However, these attributes may be dependent in the sense that two or more of them may involve similar issues. For example, the trade-off between serial and parallel operations may relate to efficiency versus reliability. In particular, parallel structures often facilitate reliability as is the case for a bus equipped with six wheels. On the other hand, parallelism is not strictly equivalent to reliability. For example, a data analysis procedure for two sets of numbers might be implemented to proceed simultaneously on a phalanx of computers, but overall system reliability may well be lower in this case than for a serial arrangement.

As alluded to above, two or more attributes may even degenerate into equivalent

*Table 2.1* Examples of the factors of intelligence

| Dimension | Realm | |
|---|---|---|
| | Natural | Artificial |
| **Purpose** | | |
| Effectiveness | Survival | Fulfilling functional requirements |
| Versatility | Wrist vs. knee | Flexible manufacturing systems |
| **Space** | | |
| Environment | Ecosphere | Factory interior |
| System | Organism | Assembly robot |
| **Structure** | | |
| Granularity | Nerve impulses vs. hormonal levels | Digital vs. analog communication |
| Network | Neural connections | Telecommunication networks |
| Hierarchy | Structure of nervous system | Industrial organizations |
| Layer | Layers of triune brain | Layering of computer languages |
| Parallelism | Blood clotting | Sequential vs. parallel computation |
| **Time** | | |
| Entropy | Evolution | Learning programs |
| Determinism | Threshold for nerve impulses | Product reliability |
| Static vs. dynamic | Homeostasis vs. growth | Regulation vs. fabrication |
| Realtime vs. batch | Neural control of muscles | Performance vs. longevity |
| Short-term vs. long-term | Endangered animal | Flight control vs. data analysis |
| **Process** | | |
| Initiative | Gas exchange in lungs | Sundial vs. clock |
| Forward vs. backward | Operant conditioning | Data-driven vs. goal-driven |
| Enhancement vs. reduction | Visual processing | Artificial visual processing |
| Feedback | Fingernail growth | Command vs. actual state |
| Centralization | Brain vs. spinal cord | Mainframes vs. microcomputers |
| Efficiency | Enzymes for reactions | Cost vs. quality |

*Table* 2.2 Examples of interfactor tradeoffs

| Dimension | Realm | |
|---|---|---|
| | Natural | Artificial |
| Space vs. time | Nutrient absorption in small intestine | Layout of factory |
| Mechanism vs. process | | |
|   Hardware vs. software | Heart valves | Mechanical vs. electronic calculator |
|   Computation vs. storage | Gene activity | Computation of trigonometric functions |

issues for specific applications. To illustrate, the degree of decentralization of production operations for a particular manufacturing plant may correspond monotonically to the degree of flexibility exhibited by the system: the decentralized control allows for rapid response to changing circumstances at the local level.

## Trade-offs

The design of an intelligent system demands the balance of conflicting attributes between two or more dimensions, as well as opposing factors within a particular dimension. The system architect must decide which attributes are most important for the system to be constructed.

An increase in one factor may be used to compensate for a decrease in another. This relationship, depicted in Figure 2.2(a), is well known to students of economics. When growing wheat, for example, it is possible to compensate for a decrease in the size of the land by using more fertilizer.

In the realm of neurophysiology, an example is found in the dendrites of excitatory synapses. The number of terminals that impinge upon post-synaptic dendrites tends to be inversely correlated with the time required to raise the post-synaptic neuron to its activating threshold. The additional synapses occupy more space, but require less time to generate a new action potential in the post-synaptic neuron.

In a similar way, the small finger-like projections, or villi, which extend from the walls of the small intestine increase the surface area available for nutrient absorption. This structural modification brings with it a corresponding decrease in the amount of time needed to absorb nutrients from ingested food.

Artificial systems as well as their biological counterparts are subject to conflicting trade-offs. In computer systems, for example, a trade-off often exists between the space available for memory or storage, and the time needed to complete a given computational procedure.

If the levels of two competing factors increase or decrease simultaneously, then system performance will be affected—for better or worse, depending on the direction of change. To return to the agricultural example: consider the case where quantities of land and fertilizer yield 100 bushels of wheat, while a doubling of both factors results in 180 bushels. The infamous law of decreasing marginal returns in the field of economics stipulates that output tends to be less than directly proportional to the levels of input resources. This decline is due to other constraining factors such as the amount of labor or machinery available.

The phenomenon of decreasing marginal returns is encountered in diverse

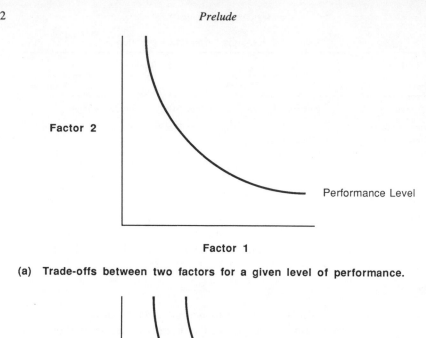

**(a)  Trade-offs between two factors for a given level of performance.**

**(b)  Increase in performance due to a rise in one or both input factors**

**Figure 2.2** Trade-off curves for two factors.

fields. In computer hardware, for example, the use of two parallel processors usually results in a speed-up of only 40 to 80 percent over that of a single unit, due to the overhead of coordinating the processors.

Only when all input factors are scaled up or down simultaneously and the system faces no overhead costs for coordination, will the level of output be commensurate with the input. In fact, enhanced utilization of existing resources may increase overall productivity, as embodied in the concept of economy of scale, where a large producer delivers goods or services at lower unit cost than a smaller counterpart.

The change in system output as a result of simultaneous changes in input factors is depicted in Figure 2.2(b). Increasing one or both factors results in increased performance. The specific application, however, will determine whether or not the output will increase in exact proportion to the increase in input.

## Summary

The design of intelligent systems may be characterized by six factors: purpose, space, structure, time, process, and efficiency. An intelligent system displays a range of behaviors that can be viewed as the means to an objective. Animate objects, in general, pursue a hierarchy of innate goals, chief among which are the survival of the self and the species. The purpose of an engineered artifact is defined by the goals of the designer. Sometimes the goals are readily apparent, as in the case of a sentry robot. In other instances the goals of the designer may be more subtle, as in the development of a docile breed of dogs. Whatever the ultimate goal of the designer, it is important to formulate a precise set of goals and to avoid being sidetracked by peripheral concerns.

Intelligent behavior is localized in some region of space. To fulfill the purpose of the system without incurring deleterious side effects, the environment must be clearly delineated and the modes of action properly implemented.

The structure of a system supports its mission. The structural concerns relate to both the physical components of the system, as well as the organizational configuration that specifies their patterns of interaction.

An intelligent object often defies the principle of entropy by becoming more ordered over time. This is achieved by exchanging order with the environment; the system must continually strive to maintain or increase its order. Often its long-term viability is enhanced by short-term sacrifice. The sacrifice might take the form of a physical penalty such as structural damage, or opportunity cost, such as the stock-piling of energy reserves.

A system interacts with its environment through a series of processes. These processes can be classified into primary functions as well as supporting activities. A primary function is dedicated to the overall objectives of the system as exemplified by reasoning procedures or the activation of limbs. The primary functions depend on supporting processes, such as intrasystem communication networks or temperature regulation mechanisms.

Any system, intelligent or otherwise, has limited resources at its disposal. For this reason, it must pursue its objectives with efficiency. The nature of the resources may be corporeal, as found in the rational deployment of materials and energy, or intangible, as in the use of efficient algorithms.

This chapter has presented an overview of the design dimensions and their relationships. These attributes are discussed in detail in Part II, and followed by a number of important interdimensional trade-offs in Part III. The dimensions are employed in Part IV to examine several case studies relating to autonomous robots, factory automation, and organizational design.

# II

# FACTORS

*This is another good occasion to point out how wrong the view is that evolution has made everything perfect . . . organisms have simply made the best of what they had and managed to survive.*[1]

Salvador Edward Luria

# 3

# Purpose

*Every detail of structure in every living creature (making some little allowance for the direct action of physical conditions) may be viewed, either as having been of special use to some ancestral form, or as being now of special use to the descendants of this form—either directly, or indirectly through the complex laws of growth.*[1]

Charles Robert Darwin

An intelligent system is defined in part by the efficacy with which it pursues some goal. Some authors would distinguish between the use of the terms *purposive* and *purposeful*. Purposive behavior denotes a physical, natural or artificial system whose objectives are assigned or inferred. In contrast, purposeful behavior applies to a system that can select its own goals and the means by which to pursue them. Such a system may be said to exhibit *will*, as exemplified by human beings.[2]

The difference in the underlying concepts may be central to certain applications. On the other hand, the associated nomenclature—purposive versus purposeful—is not generally recognized in the literature. In this book the two terms will be used synonymously, and the internal or external origin of a system's objectives will be clarified only where critical.

We characterize a purposive system with the following definition:[3]

- An action **a** of a system has the *purpose* or *goal* **g** if the following conditions apply:

    -The system may choose whether or not to perform action **a**.

    -The system is aware that executing **a** results in, or increases the likelihood of, attaining **g**.

We take the view that any artifact that fulfills a set of functional requirements is a purposive object; an example is a manufacturing system and its components.

The goals of an engineered system are assigned externally, whether from its creator or another engineered device. These specifications may be classified into *functional requirements* that define the purpose of the system, and *constraints* that limit the space of secondary characteristics. The functional requirements are often associated with a tolerance band, which defines the allowable variation in the underlying parameter.

To illustrate, the functional requirements of an engine may be to provide a power output of $100 \pm 5$ kilowatts at a rotational speed of $4000 \pm 30$ revolutions per minute. The design problem may also be characterized by constraints such as a lower limit on reliability and upper limits on weight and size.

To pursue another example, the purpose of a flexible factory might be embodied in the following objectives:

$G_1$ = Manufacture many products simultaneously.
$G_2$ = Accommodate a wide range of lot sizes.

The quantification of these objectives results in the *functional requirements* or *performance levels* of a design implementation. A particular system may offer the following functional requirements:

$F_1$ = Manufacture from 1 to 1000 products simultaneously.
$F_2$ = Accommodate lot sizes from 1 to 1,000,000.

Suppose, however, that a second system can produce up to 2000 products simultaneously and accommodate the same range of lot sizes. By virtue of the enhanced capability, the second factory may be considered more intelligent than the first. Hence the superiority of performance is tantamount to increased intelligence.

The ability of an engineered system depends on the fulfillment of its functional requirements as well as secondary constraints. These constraints delimit the acceptable bounds of secondary characteristics. The constraints on a flexible factory might be as follows:

$C_1$ = The system must reconfigure for a new product within 1 second.
$C_2$ = Average inventory level should remain below 1 day.
$C_3$ = Construction cost should be under $100,000,000.
$C_4$ = Operating cost should be less than 10 percent of the value added.

The functional requirements and constraints collectively define the design task in terms of measurable quantities.

## Effectiveness

*Effectiveness* is a measure of the degree to which a system attains its goals. In a complex, dynamic world, the system may need to adapt to unforeseen circumstances and persevere in spite of external as well as internal disturbances. In this context, the system architect should consider the following parameters of capability:

- *Performance.* A purposive system must satisfy its specifications in terms of functional requirements and constraints.
- *Versatility.* A rich behavior space allows the system to respond to new circumstances. The flexibility of behavior may be implemented by providing for a large potential set of capabilities, as well as ease of system reconfiguration to implement alternative patterns of behavior as needed. To illustrate, an amphibious craft may operate as a land vehicle in one configuration and as a boat in another.

In an imperfect world governed by friction, entropy, and randomness, no system will perform at peak capacity for an indefinite period. Given the inevitability of lapses in performance, a set of constraints is often specified for minimal levels of acceptable operation.

- *Reliability*. This factor indicates the extent to which a system can operate without failing. A commonly used metric is the mean time between failures.
- *Availability*. A system may operate for extended periods without mishap, but may be of little use if it cannot be repaired in a timely fashion. Hence availability measures the fraction of operating time to the total time over an extended period.
- *Maintainability*. This parameter indicates the ease of keeping a system in operation through preventive maintenance such as regular cleaning.
- *Serviceability*. A system may work effectively over extended periods through a minimal maintenance schedule, yet be prohibitively expensive to repair. Serviceability is a metric of the cost of system repair due to breakdowns.

Of the factors above, versatility may be the most critical aspect: a free device that always works while performing irrelevant tasks is of limited use. The next subsection delves further into this topic.

## Versatility

Versatility refers to the portfolio of capabilities exhibited by a system. The appropriate level of versatility is defined by the purpose of the system as well as the nature of its environment and the resources available for initial design.

In a static universe, it may be possible to determine the relationships between a system and its environs, thereby allowing for a relatively simple determination of requisite system capabilities. In a probabilistic universe whose only constant is change, the system goals may evolve over time, as may the environment. These factors complicate the determination of the optimal degree of versatility.

A low level of versatility is correlated with tractability, or ease of system design and maintenance. On the other hand, high versatility implies flexibility of system response and the ability to adapt rapidly to changes in requirements or in the environment.

Flexible systems often possess complex architectures which can be difficult to understand and master. Given the expanded repertoire of capabilities and the attendant complexity, it may be difficult to identify the conditions under which one function should be selected over another.

In the human body, neurons are a good example of a flexible system. Each of the various components of a neuron is quite versatile, possessing the ability to perform a variety of functions. The dendrites of a neuron can also perform several functions in addition to their primary task as receptors (see Figure 3.1). In certain cases dendrites can act as effectors, transmitting electrical impulses to other dendrites. These junctions are known as dendrodendritic synapses.

In fact, almost any part of the neuron can adapt and learn to perform new functions if required by circumstance. For instance, if part of an axon is destroyed,

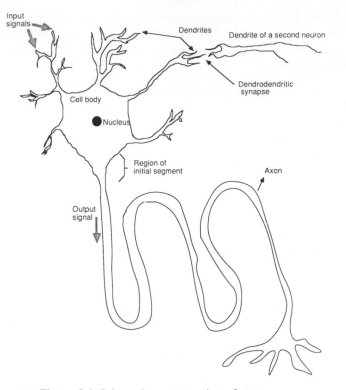

**Figure 3.1** Schematic representation of a neuron.

a region of dendrites may take over its role and assume responsibility for impulse conduction. The flexibility of neurons enables them to adapt and function successfully in the face of changing conditions.

The study of anatomy provides further examples of the degrees of versatility found in various systems. Human joints range from being very flexible to very tractable, depending on their design and intended function. Both the knee and the wrist are complex joints; each contains several different bone articulations, as well as cartilage, tendon, and ligament connections. However, the knee is highly constrained. Essentially, only one type of motion is allowed—either flexion or extension—and rotation about the knee is basically impossible. In contrast, the wrist is highly flexible. The joint allows for full rotation of the hand, in addition to flexion and extension.

The trade-off between versatility and ease of use also arises in the selection of a software utility to develop an application package. In the development of a conventional software package, a specialized tool such as a financial modeling package or a simulation language may be used to develop the application rapidly in lieu of a general purpose language such as Cobol, C, or Pascal.

Unfortunately, this approach faces the following disadvantages. First, the user has little or no capability to incorporate novel aspects of the application which were not already built into the package. For example, a user cannot automatically generate statistics on multiple simulation runs of a financial model if the modeling package does not already incorporate such features. The second major disadvantage of specialized languages lies in their isolation. No software module is an island unto itself: it is one piece in the tapestry of organizational processes. But a module written in a specialized language may offer few or no mechanisms for interfacing with other software packages.

This trade-off between ease and versatility applies equally well to advanced systems incorporating artificial intelligence. Using a specialized tool such as a *shell* or *development environment,* a prototype expert system might be developed within a matter of days or weeks rather than the five workyears which often characterize the minimum amount of effort needed to develop a sizeable expert system from scratch. On the other hand, use of a general-purpose language such as Prolog or Lisp offers flexibility and improved mechanisms for system integration.

The flexibility of a production system relates to its range of behaviors. The increasing market segmentation of consumer tastes implies reduced production lot sizes and thereby the need to enhance flexibility even at the cost of decreasing tractability.

A typical flexible manufacturing cell is depicted in Figure 3.2. A robotic manipulator accepts incoming parts on a conveyor belt and transfers them to various machine tools depending on the need for performing milling, turning, or deburring

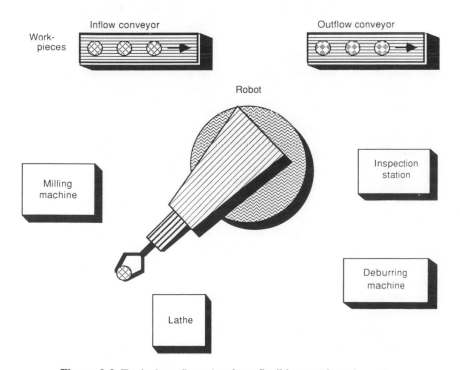

**Figure 3.2** Typical configuration for a flexible manufacturing cell.

operations. After the appropriate operations, each workpiece is checked at an inspection station and finally placed on the outgoing conveyor belt for transport to the subsequent production cell. The versatility of the system ensures its long-term viability in the midst of a dynamic industrial environment.

## Knowledge, Information, and Data

For a system to interact intelligently with a changing environment, it must be aware of its environs and perhaps even of its own condition; such awareness may take the form of knowledge, information, or data. Knowledge refers to an awareness of the state of the world or of procedures for attaining goals. Information is a subset of knowledge pertaining only to the state of the world. Data is a quantity that can be transformed into information by the use of knowledge. We summarize these notions in the following definition:

- A system has *knowledge* **K** if **K** is a fact relating to the state of the world (whether inside or outside the system), or if **K** can be used to attain some goal.
- A system has *information* **I** if **I** is a type of knowledge pertaining to the state of the world.
- A system has *data* **D** if it can use its store of knowledge to transform **D** into information.

In everyday speech, we associate information with the value of a communication. A medical journal is likely to convey more information to a physician than to a carpenter, for example.

When we speak of "value" there is a notion of utility, implicit if not explicit, with respect to some purpose or objective. The value of a communication depends on the recipient and his goals. A communication that takes a system closer to its goal conveys information; one that takes it further away conveys disinformation. Since we live in a nondeterministic world, we may generalize the idea of "moving closer" to a goal by saying "more likely to attain" the goal. Hence, information may be defined in terms of the likelihood of attaining a set of goals. (A quantitative formulation of this concept is presented in Appendix E.)

Information is required to attain goals in the face of changing environmental conditions. Hence, loss of information can result in the degradation of system performance. An example lies in the genesis of instability: a control system of first or second order, which is inherently stable as a continuous-time system, may become unstable as a discrete-time system. We may attribute this metamorphosis to the partial loss of information resulting from sampling information at discrete intervals rather than continuously.

Another class of instability occurs in various nonlinear elements such as backlash components. Here, the loss of information may be traced to lack of knowledge about the relative displacement of interfacing elements within the dead zone. It is clear, however, that not all nonlinear elements result in instability. An example is found in saturation effects which produce limit cycles exhibiting stable behavior.

## Summary

The design of an intelligent system begins with its purpose. The objectives of a system may be explicitly assigned by another intelligent agent, or implicitly conferred by nature.

- The performance of an intelligent system is defined in terms of its purpose. The purpose may be explicit or implicit.
- The purpose of an engineered system is specified quantitatively in terms of functional requirements, and is qualified by constraints relating to secondary characteristics.
- System effectiveness may be evaluated in terms of performance and versatility.
- System constraints often include reliability, availability, maintainability, and serviceability.
- Knowledge refers to an awareness of the world or to procedures for attaining a set of goals. One component of knowledge is information, relating to the state of the world, both in terms of the external environment and the internal condition of the system. Data is any quantity that can be transformed into information through the use of knowledge.

# 4

## Space

*In turbulent times, the first task of management is to make sure of the institution's capacity for survival, to make sure of its structural strength and soundness, of its capacity to survive a blow, to adapt to sudden change, and to avail itself of new opportunities.*[1]

Peter Ferdinand Drucker

### Localization of Intelligence

Intelligent behavior can be localized in space where it expresses itself through some physical embodiment. Although the extent of the spatial localization is often readily apparent, at other times it remains open to debate. In describing intelligent behavior, it is useful to refer to some physical component or region of space even though the precise boundaries may be somewhat arbitrary.

The dichotomy between a system and its environment—or its sibling concept, nature vs. nurture—pertains to the relationship between an object and its surroundings. This delineation defines the richness of interactions across the interface as well as the complexity within each realm.

A system must adapt its structure and behavior to thrive within a changing environment. The requisite complexity of a viable system is therefore determined by the complexity of the environment.[2] For example, if the environment can exhibit ten levels of temperature, the system must possess at least ten states or conditions to maintain itself at some predetermined condition.

For the purpose of discussion, we may distinguish the actual internal complexity of the system—as represented by the system's structure and information processing modes—from its apparent complexity—as reflected in its observable behavior. Moreover, both the environment and the system may be partitioned into two major spheres: the interface and the hidden components.[3] This arrangement is depicted in Figure 4.1.

The interface of the environment defines its direct interactions with the system. The larger, latent component of the universe that envelops the system interacts with it only indirectly, through the interface. For example, a robot's operation may be affected directly by the ambient humidity in its immediate locale, but is largely independent of the atmospheric conditions on the other side of town.

In a similar way, the interface of a system relates to observable characteristics,

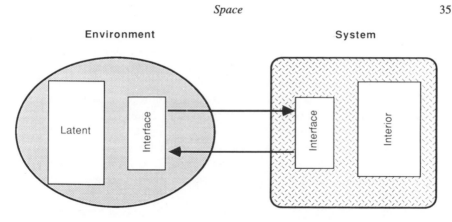

**Figure 4.1** Partitioning of system and environment into the apparent and hidden components.

whether in terms of physical or intangible qualities. In contrast, the interior aspect of the system—its hidden components—refers to the mechanisms and processing requirements needed to support the interface. For example, the external body of a robot and its perceptible behavior define the interface. The body might be supported by an internal network of beams, and the apparent behavior moderated by a set of software procedures; these collectively constitute the interior of the system.

## Environment

The environment may exhibit different types of interfaces to the system. As with a system, a topology of environmental conditions may be defined by attributes such as space, time, initiative, and others. Some examples, in order of behavioral complexity, are given below:[4]

- *Placid-randomized.* In this environmental texture, the positive components (e.g., food or resources) and negative elements (e.g., obstacles or malicious factors) are randomly distributed over space and unchanging over time. Due to the random nature of the environment, the system's adaptive response will show little differentiation between its short-term tactics and long-term strategies.
- *Placid-cluttered.* Here the positive and negative elements are still static over time. However, they are clumped together in structured ways rather than scattered randomly. A system is unlikely to flourish in this context with a simplistic strategy; it must develop an overall plan for negotiating the large-scale terrain in addition to tactics for coping with specific situations in the immediate vicinity. For example, it must learn to gravitate toward areas rich in resources while avoiding dangerous elements.
- *Disturbed-reactive.* This environment is similar to the previous one, with the added complexity of competition. Here, the system pursues its goals in competition with other agents of similar ends and means. Not only does the system

maintain itself against an indifferent environment, it must consider goals and activities of other intelligent agents. Hence, in addition to strategies and tactical plans, the system must develop a repertoire of mediate-term operating plans. The system that prospers will be one that has effectively resolved issues such as, "When is cooperation better than competition?" and "With whom should I cooperate, and against whom should I compete?"

- *Turbulent.* Here, a system competes against agents of similar constitution within a dynamic context. In other words, the nature of the environment itself appears to be changing over time. For a robot in an automated factory, such unexpected behavior may result from the collapse of a gangway or the introduction of a new machine tool. For an industrial organization, uncertainty may arise from changing social values or a technological breakthrough from an unexpected quarter. In the midst of such uncertainty, the system has little recourse but to draw on its past and present in an attempt to anticipate possible futures, remaining flexible to accommodate changes.

It is clear that the structure and behavior of an intelligent system is defined not only by its goals but also by its surroundings. The more complex the environment, the more elaborate must be the requisite system interface and internal structure.

In the societal sphere, the viability of an organization depends on the compatibility between a system's objectives and the needs of the larger society in which it operates.[5] An organization must attempt to resolve and meet the conflicting demands of society by channeling its stock of resources and distinctive competence into productive output. If the organization succeeds in this effort, it is entrusted with more resources and thereby flourishes; otherwise the system withers.

Environment plays a key role in defining a system's structure. A study of ten companies in three industries illustrates this relationship.[6] The plastics industry, for example, is characterized by frequent technological advances and market turbulence. Because this environment is so diverse, the companies within this industry differ greatly from one another in both structure and behavior. The consumer food industry is characterized by a moderate rate of change which in turn is reflected in the mediate level of differentiation within the industrial organizations. The standardized container industry has the most stable technologies and markets among the three industries studied; here the various groups are characterized by homogeneity. Hence, successful organizations are those that exhibit an effective balance of differentiation (decentralization) and integration (centralization) in line with the demands of their environs.

An organization of human dimension must contend not only with the physical environment, but the artificial arenas of economics, politics, society, and technology. *Economic* considerations arise because organizations, like natural entities, must take in nourishment to survive. In human affairs this nourishment takes the form of financial currency: schools charge tuition, companies issue invoices, charities seek donations, and governments levy taxes.

*Political* factors arise from the need to deal with governments and their constituencies. In the 1980's, many American companies were prevented from selling their wares to Soviet customers. The barrier was political rather than financial, for

U.S. policy makers decided to block the flow of technologies and devices that could be used for military ends by a potential adversary.

The *social* influence on an organization springs from the fact that its goals and tactics must conform to external conditions. One category of social constraints relates to values and morals. If evil seldom triumphs on television, this is a result of broadcasters' policies, rather than the belief that reality so often favors the good. If mass advertising among physicians and lawyers is still sporadic and limited, it is not due to the ineffectiveness of publicity but rather to a sense of propriety, both within the professions and among the potential clients. Consumer companies such as Nestlé and Coca-Cola have been attacked for the intensity of marketing efforts directed to the impressionable poor in third world countries, those who are cajoled into spending a week's income on a soft drink or infant formula that is diluted beyond nutritional value.

When the television series *Star Trek* first appeared in the mid-1960's, it was inevitable that all the chief roles would be filled by males. Two decades later, the new generation of stellar trekkers included female chiefs for medicine and even security, although the top three officers of the good starship *Enterprise* still remained male. This new dimension in the composition of the crew reflects the ongoing change in society's perception of the role of women.

Another important societal consideration relates to demographics. The baby boom following the Second World War constituted a major market segment, certainly in numbers if not in absolute purchasing power. The advertising thrust and even the society at large seemed to focus on youth, energy, and vitality. But now the baby boomers are adults, and the teenage models have given way to middle-aged women. As a result, the focus of society has aged along with them. A new emerging market centers on the senior population consisting of individuals aged 60 and above. In absolute numbers as well as relative figures, the "golden age" group constitutes a segment of the population of unprecedented significance in the history of the world. It is no surprise that gerontologists are in increasing demand by both scientific and commercial enterprises.

In today's society, *technology* sets the pace of change. Styles may change over the course of years and values may evolve over decades, but technologies may change overnight. If people are more informed about world affairs than in the last century, it can be explained better as a result of communications technology rather than an increase in curiosity. By the same token, people live longer than they used to because of advances in medical technology and sanitary practices rather than an enhanced zest for life.

The importance of effective interfacing with the environment is also found in organizational behavior. An organization can modify its operational arena by constructing a *negotiated environment*.[7] This may be achieved through external vehicles such as lobbies, industry agreements, standards, and customer contracts.

Molding the environment is a tactic employed by government agencies as well as private institutions. The National Aeronautics and Space Administration, for example, must be mindful that its project portfolio includes some activities that arrest the public vision and thereby ensure support for its operating budget. The Voyager spacecrafts sent as envoys to the stars are of little value to Earthkind after their departure from the solar system, but this program galvanizes the human imagination more than the scientific benefits resulting from entering into Jupiter's orbit or even hurtling into Saturn. For similar reasons, private corporations sponsor the arts and conduct public relations programs.

## System

A system may be partitioned along two key dimensions relating to corporeity and cognizance. Each of these dimensions can in turn be divided into two categories.

The degree of corporeity may be identified in terms of the physical aspect relating to tangible objects, and the abstract aspect pertaining to intangibles. On the other hand, the level of awareness may be partitioned into the object level, relating to the basic or observed system; and into the metalevel, denoting the observer of the basic level. The object level performs its activities mindlessly, without explicit knowledge of ultimate goals. The metalevel monitors the object level, determining how well the goals are met, and guiding the object-level actions as required.[8] The interactions of these dimensions result in the four components shown in Figure 4.2.

The basic component of an intelligent system is the hardware or mechanism, as

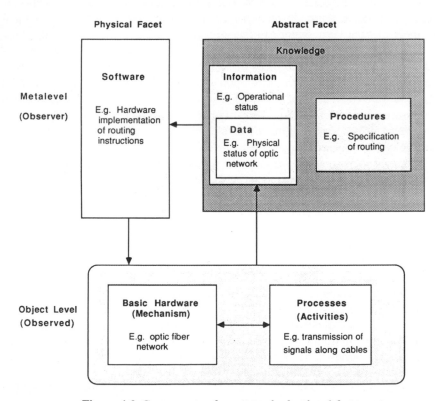

**Figure 4.2** Components of a system, by level and facet.

exemplified by an optic fiber network. This mechanism harbors a series of processes that may consist of deliberate activities such as the transmission of signals along the optic cables, or natural phenomena such as contamination from the environment.

The mechanism and processes are monitored at the metalevel. The knowledge component consists of data such as the status of the optic network, and procedures such as the specifications for message routing or instructions for self-repair. This knowledge is encoded in some language or code, and implemented on some type of hardware. The specific implementation may take a variety of forms, from phenomena such as chemical structures in biological systems to magnetic polarities in artificial recording devices.

This basic framework for systems may be refined or specialized for specific situations. To illustrate, the observing level may be decomposed into first-order and second-order metalevels. A sample second-order metalevel object is a procedure that regulates the operation of basic procedures, or even modifies them.

On the other hand, a physical system that has no intelligent characteristics may be regarded as having a null metalevel. For example, pebbles on a beach belong to this class of systems.

We note in passing that the term *software,* as used in this chapter, is more restrictive than is found in the general literature which uses the term to denote anything other than the basic hardware: physical processes and knowledge as well as machine implementations. As discussed above, however, it is possible to separate these three concepts.

The utility of this separation will become clearer in succeeding sections such as Chapter 10, which addresses the interplay between mechanism and process. For instance, when computer designers speak of the trade-off between hardware and "software," they are actually referring to the dichotomy between hardware and process. In this context, the software is the specific set of instructions that implement the conceptual procedures required to activate a process. Hence, the same observable process may be implemented using a different set of instructions or even procedures.

## Internal Complexity

How does complexity in the environment affect complexity required in the system? A trade-off exists between the extent of the system's preprograming and, conversely, the ability of the system to monitor and respond to changing or unexpected initial conditions in its working habitat. In order to properly adapt to the environment, a system must possess at least as many internal states as the number its environment can assume.

An example of this is the ability of animals to regulate their body temperatures. If the environmental temperature fluctuates among three possible states—cold, warm, and hot—then the organism must also possess at least three internal states in order to maintain its body temperature despite changing external conditions. When the environment is much colder, the animal must produce heat to compensate for that which is lost, and thereby avoid a drop in its body temperature. If the environmental temperature is significantly higher than the animal's body temperature, the animal must develop a cooling system to avoid overheating itself. And in the event that the external temperature is approximately equal to the animal's body temperature, no special heating or cooling systems are necessary. The organism therefore has three internal states: heat-producing, cooling, and neutral.

Skin pigmentation in humans is also a result of adapting to environmental conditions. Natives of areas where the sun's rays are more intense tend to have darker skin than those in regions where the heat is less intense and more shade is available. Thus, the system is influenced toward phylogenetic change by the environment. In the case of pigmentation, the change toward darker or lighter skin occurs not within one individual's lifetime, but over a much longer period.[9]

The structure of an organization may also adapt in the face of a changing environment. A case in point is the Matsushita Electric Company, producer of brand names such as Panasonic and Technics. The firm was founded in 1918 by Konosuke Matsushita and is now one of the largest industrial firms in the world. Following the lessons of the Du Pont and General Motors companies in the United States, Matsushita adopted a divisional structure in the mid-1930's, even though there were only 1600 employees at that time.[10] Each division, having responsibility for a specific line of products, could follow its own strategic plans. Divisional activities were subject only to relatively loose constraints defined explicitly by the mother corporation or implicitly by the sister divisions.

In 1943, the firm was slightly re-organized into product groups in which division managers reported vertically to the president and horizontally to their respective group vice presidents. This aspect of decentralization allowed for better cohesion within related families of products. This also represented the beginnings of matrix organization, which would gain popularity among U.S. companies a decade later.

In response to the recession and general disarray following World War II, Matsushita restructured the company once again, this time into a highly centralized format to facilitate a focused strategy. As economic conditions gradually improved, the firm was decentralized yet again after 1953 to respond flexibly to increasing competition.

The second half of the 1950's was a period of dramatic growth and international diversification, a period in which the firm recentralized once more. The decade of the sixties was a confusing time economically, and decentralization was in order. A period of centralization followed, along with the oil crisis and economic stagnation of the mid-1970's.[11]

This corporate example reflects the dynamic relationships between a system and its environment. To fulfill its goals, the system must restructure its internal configuration to deal effectively with the environs.

## Interface Complexity

This characteristic relates to the source or imputation of apparent complexity in system behavior. In other words, what are the relative contributions of complexity in the system versus the environment? At times the apparent richness in system behavior can be attributed more to the environment than to the system itself.

A well-known example lies in the seemingly complex path of a single ant across a beach molded by wind and water. A sketch of its path, complete with detours around pebbles and sand dunes, resembles a connection of linear trajectories. They are not random, however, because the ant has some general sense of its homeward direction.

When viewed in isolation, the ant's path seems quite complex, full of inexplicable twists and turns. Yet this complexity results from that found in the environment. The ant's roundabout course results from the detours it is forced to make to avoid natural obstacles in its path. Thus, an ant's behavior is comparatively simple. The apparent complexity of its behavior over time is essentially a reflection of the complexity encoded into the ant's environment.[12]

We can also find examples where a system is much more complex than it appears to be at first. The fact that every cell in a multicellular organism possesses the same genetic code suggests that these organisms are exceedingly simple. However, this is misleading; the structural organization of the organism is in fact very complex, and this complexity arises from environmental influences, including those of one cell on another. The cell's environment activates certain genes and inhibits others, allowing cell differentiation to proceed and form the foundations for the system's architecture. In this manner different cells, tissues, and organs are produced, ultimately creating a unique individual.

Even very simple machines can appear to act intelligently when the environment assigns complexity to the simplest of tasks. By connecting various mechanical devices to each other, we can construct machines that seem to "think." In reality, however, the various components are merely interacting with one another and with the environment in simple ways.

This phenomenon of apparent intelligence arising from mindless activity can be highlighted through a series of thought experiments using simplified vehicles. The following discussion is adapted from an essay by the neuroscientist Valentino Braitenberg.[13]

The qualities we recognize as aggression and cowardice can arise through alternate arrangements in a simple control mechanism. We begin by constructing a vehicle with two sensors, one connected to a motor on the right side of the vehicle, the other on the left. The output of each sensor varies with the intensity of the incident stimuli and thereby affects the speed of the motor. The way in which we connect the sensors to the motors will also affect the behavior of the vehicle. Consider the case where each sensor is connected respectively to the motor located on the same side (see Figure 4.3). If the stimulus is located on one side of the vehicle, the sensor on that side will relay that information to its motor, and the motor on that side will turn faster. Consequently, the vehicle will turn away from that stimulus.

In the second case, where each sensor is attached to the motor on the opposite side of the vehicle, the direction of motion will be different. In this situation, the vehicle will approach the source of the stimulus, instead of avoiding it. Since the

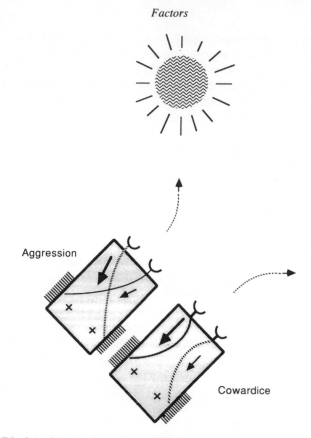

**Figure 4.3** Display of aggressive and cowardly behaviors. On each vehicle, the differential levels of stimulation on the two sensors and their respective wheels produces motion toward or away from the light source. [After Braitenberg (1984), p. 8.]

first vehicle constantly tries to avoid its stimulus, one could claim that it exhibits "cowardly" tendencies. On the other hand, since the second vehicle tends to seek out its stimulus, it exhibits aggressive behavior. Thus machines with very simple connections can produce behavior that is reminiscent of anthropomorphic tendencies.

The Darwinian principle of survival of the fittest can apply as well to inanimate beings as it does to living creatures. By modifying the connections among existing components and adding more components, the creator of the vehicles can produce a strain of machines that incorporate the effects of experience into their hardware. This may be achieved, at least in part, by using hardware whose properties are history-dependent. For example, we can construct a vehicle that steers clear of the edges of the table top, thus managing to avoid falling off and crashing. This vehicle may in fact have very simple wiring so that, for example, it moves only in circles. But over time, Darwinian evolution will favor those vehicles that avoid the edge, because the rest will wander to the sides, fall off, and destroy themselves. The vehicle's pattern of movement need not be confined to a circle; it may even have a complex pattern, encoded within the hardware, which the vehicle keeps repeating. If the characteristics of surviving vehicles determine—at least in part—the properties of their successors, Darwinian evolution is effectively illustrated. Eventually the table top will be populated by vehicles possessing some sort of "instinct" that allows them to travel in safe paths.

The appearance of high-level concepts can also be attributed to evolutionary mechanisms. We can construct vehicles that appear to associate objects with concepts, such as danger. For example, suppose that the table supporting our vehicles contains obelisks that can inflict irreparable damage to the machines. Through Darwinian evolution, surviving vehicles would manifest obelisk-avoidance behavior. This gives rise to the apparent association of the concept of danger with obelisks. Further, suppose there is a certain type of obelisks stationed near the edges of the table; then the vehicles will seem to associate the "fear of falling" with those objects. Although the vehicle avoids the objects via a combination of sensory and motor functions, an observer can impute to that behavior the cognitive function of associating the objects with danger.[14]

Intelligent behavior can spring from unintelligent components. Further, seemingly rational behavior and anthropomorphic attributes can arise even in the absence of an explicit objective.

## Summary

A system operates within a larger context. It is sustained by the environment and must act in accord with the constraints of its milieu. A special advantage of an intelligent agent is its ability, at times, to modify the environment to better fulfill its own objectives.

- The character of a system depends on the nature of its environment: the more complex the environment, the greater the requisite organization of a viable system.
- The apparent sophistication of a system's behavior can be resolved into active and passive components. The active sophistication springs from the richness of its internal organization, while the passive component arises from the complexity of the environment.
- A human organization must contend not only with its physical environs, but with the synthetic dimension of economic, political, social, and technological factors.
- A system may be partitioned along two interacting axes: corporeity in terms of physical and abstract components, as well as cognizance in terms of the scope of awareness.
- Intelligent behavior can result from simple components. Further, behavior of apparent rationality can arise even where explicit goals are lacking.

# 5

# Structure

*Brain and mind alike consist of simple elements, sensory and motor.*[1]
William James

*I'll call "Society of Mind" this scheme in which each mind is made of many smaller processes. These we'll call agents. Each mental agent by itself can only do some simple thing that needs no mind or thought at all. Yet when we join these agents in societies—in certain very special ways—this leads to true intelligence.*[2]
Marvin Minsky

## Purpose versus Structure

The structure of a design should support the intended objectives of the referent system. More specifically, structure and purpose relate to the correspondence between a system's architecture and its specified function. In some fields, this relationship is better known as the interplay of form versus function.

All intelligent behavior as we know it must be implemented on some hardware. For this reason, the structure of a system is a critical determinant of its abilities. According to Norbert Wiener, the father of cybernetics, a machine that duplicates human physiology would embody the intellectual capacity of a human being.[3] In fact, it seems plausible that an engineered system that precisely duplicates a person using identical biological, chemical, and physical processes would be indistinguishable from a person.

The four specialized types of human body tissue—epithelium, connective tissue, muscle, and nervous tissue—exemplify how systems develop to serve different purposes. Each tissue possesses distinct features that make it especially suited to carry out its intended functions. Also, within each type of tissue, further differentiation occurs to increase the tissue's efficiency in its given environment.

The category of connective tissue, for example—which includes blood, connective tissue proper, cartilage, and bone—possesses unique qualities that facilitate its primary function: to provide soft "packing." One of these qualities is an abundance of extracellular matrix substance. However, the extracellular matrix composition itself varies. Red and white blood cells are suspended in an aqueous environment especially suited for the transport of soluble substances, which is the main function of blood. On the other hand, bone extracellular matrix is composed of extremely hard fibers reinforced with calcium and lime. These extracellular components endow bones with the strength to support our skeleton.[4]

The dependence of structure on purpose is also found in enzyme organization. Structural relationships in a cell's protoplasm help maintain the orderly conduction of cellular processes. Enzymes can be organized into multi-enzyme aggregates that increase the overall efficiency of processes which are catalyzed. For example, the synthesis of lipids from fatty acids can be catalyzed by multi-enzyme complexes known as fatty acid synthases. Although the constituent enzymes of such complexes can exist and function separately in one-celled organisms, fatty acid synthases cannot be subdivided without some corresponding loss of activity in yeast, mammals, and birds.

This grouping of several enzymes with related functions into a separate structural unit increases the efficiency of each in the achievement of its individual purpose. Placing related enzymes together ensures that the synthetic reaction will reduce processing time, because the products of one reaction will be readily available for the next reaction in the sequence. Moreover, the compartmentalization of reactants, products, and enzymes also ensures that competing reactions will not gain access to and monopolize needed reagents; hence, the collective efficiency of the enzymes is maximized.[5]

The variation among different neuron types provides another interesting example of the interplay between purpose and structure. The size and shape of each neuron depends to a large extent on its task. Although almost all of the cells in the vertebrate nervous system share the same basic features, individual neurons exhibit modifications of this basic cellular plan that enable them to carry out their specialized tasks with increased efficiency.

For example, bipolar neurons possess two processes, which usually extend in opposite directions from the cell body. The distal end of each process gives rise to many branches. At one end of these branches are the dendrites that sense impulses; at the opposite end are the axonal branches which transmit signals. Because of this structure two long processes with branches at each end the bipolar neurons are especially suited for conveying information, sometimes over long distances, from peripheral receptors to integrating centers.

In contrast, the Purkinje cells in the cerebral cortex, as well as other sensory neurons, possess highly arborized networks of dendrites. The Purkinje cells appear to be the primary integrative centers in the cortex; thus they must receive and assimilate a large number of inputs arriving from various sources. Consequently, the dendrites of these neurons are highly branched structures which extend in all directions. This type of generalized and far-reaching network of dendrites seems ideal for a neuron whose main function is the integration of many inputs.

Thus, even though neurons themselves are specialized types of cells designed to carry out the task of information-processing, different subclasses of neurons exist that exhibit even further differentiation, enabling them to carry out their tasks quickly and efficiently.

An evolutionary example of structure being determined by function can be found in certain animals. Ducks, for example, which spend most of their lives in aquatic environments, possess webbing between the digits of their feet. By providing more surface area for water resistance, the webbed feet are suited for swimming. Humans and other terrestrial animals have lost the webbing and developed separate digits which are useful for life on land. Since terrestrial animals no longer have to contend with water resistance when moving in their environment, webbed digits are unnecessary; in fact, they would impair their manual dexterity.

Evolution is also responsible for fashioning different bone structures for birds and humans. The structure of bird wings differs from that of the human arm due to the distinct requirements demanded of each limb. Since birds spend a good part of their lives flying in the air, they developed as many adaptations as possible to reduce their weight. One of these is the hollow structure of their bones. Human bones, on the other hand, are filled with bone marrow as well as trabeculae or bony latticework. Further, the bones in the human hand and digits are arranged linearly, with a simple tubular construction. In contrast, the bones at the distal end of bird wings form V-shaped struts for further reinforcement. With the addition of these features, the avian bones are particularly suited to their tasks of carrying the long feathers and providing a stiff axis for the distal portion of the wing.[6]

In the realm of artificial creations, an autonomous mobile robot can move about using wheels, treads, or legs. Wheels are efficient for transforming motive power in translational motion. They work at high speeds and offer maneuverability even to the point of having zero turning radius. However, they are of limited use in rough terrain or for negotiating obstacles. On the other hand, treaded tracks such as those found on military vehicles provide traction through a large contact surface and can support heavy payloads; they can also negotiate rough terrain. However, tracks are slow, cumbersome, and energy-inefficient. Of these methods, legs provide the greatest dexterity and mobility, including the capability to ascend steep grades. On the other hand, our limited understanding of legged locomotion permits engineered systems to support only limited payloads. For these reasons, the structure of a mobility subsystem for a robot is dictated by its functional requirements and constraints.

## Modularity

*Modularity* refers to the decomposition of system structure into smaller units. The lowest level of decomposition may be referred to as *granularity*, which defines the size of the basic units or chunks that constitute a system.

The appropriate level of granularity is defined by the prospective functions of a system. Fine granularity allows for precise control but entails a high degree of coordination. In contrast, coarse chunking allows for ease of manipulation at the expense of fine-tuning capabilities.

For example, a set of toy building blocks that contains only large cubes will facilitate the construction of a large doll house. On the other hand, small details in the doll house such as attic windows may not be depictable using the blocks. In a similar way, a set of machine instructions for a computer that can address memory only at intervals of 32-bit words will limit the ability of a programmer to manipulate data at sub-word levels.

The granularity of a system can be classified according to its *continuity* as either discrete or continuous. Although we often view discreteness and continuity as qualitatively different aspects, it is also possible to regard continuity as a limiting case in which the size of discrete chunks goes to zero.

The firing of neurons as an all-or-nothing activity illustrates the notion of a discrete phenomenon: we can talk about one or six neurons firing at once. When dealing with hormone levels in the body, on the other hand, it is more appropriate to view the amount of hormone as a continuous variable. Thus, the firing of neuronal impulses is discrete, whereas hormonal levels are continuous.

The transmission of neural signals along an axon may be either discrete or continuous, as suggested by Figure 5.1. An action potential or signal moves along an axon by inducing local current in the form of a flow of ions across a cell membrane. Whether the propagation occurs discretely or continuously depends on the structure of the axon serving as the transmission "wire." Some axon fibers have a plain exterior while others are surrounded by a coating of myelin, a soft white fatty material. The myelin coating is formed by special cells that wrap themselves around the axon at regular intervals, much like tape around a garden hose.

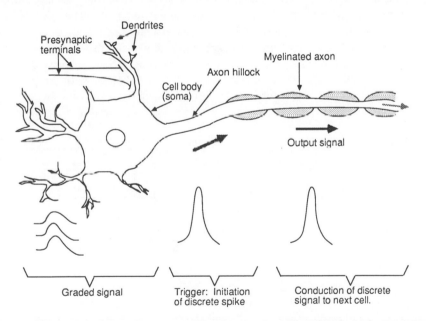

**Figure 5.1** Initiation of a signal in a motor neuron. Incoming signals at the presynaptic terminals are continuous or graded; these are transmitted as graded inputs to the dendrites, which may use the signals as positive (excitatory) potentials or negative (inhibitory) potentials. An action potential is created at the axon hillock and travels down the axon in discrete steps. (After Thompson, *Foundations of Physiological Psychology*, Harper and Row, New York, 1967.)

In fibers without a myelin coating, each action potential induces current flow in adjacent areas of the membrane, depolarizing the membrane to its threshold potential; at this point the process begins anew. This sequence of events—action potential, local current, depolarization—repeats itself continuously, as the impulse travels along the neuronal membrane (see Fig. 5.2). The generation of each new action potential depends on the electrochemical characteristics at the new site, which usually tend to be identical to those of the old. Hence, new action potentials are fresh copies of their predecessors.

In unmyelinated axons, the speed of signal transmission depends on the size of the axons. For mammalian nerve fibers, the speed increases with fiber thickness, from about 0.45 meters per second for an axon of 0.5 micrometer diameter, to about 2 meters per second in fibers of 1 micrometer.[7]

In myelinated fibers, the fatty coating of myelin prevents current flow between the interior and the exterior of the axon. Thus, action potentials are not initiated in the parts of the cell membrane covered with myelin. Instead, discrete action potentials occur only at the nodes of Ranvier, places on the axon located between myelin coats. This mode of signal propagation is shown in Figure 5.3.

Discrete and continuous conduction of action potentials offer both advantages and disadvantages. Discrete conduction is characterized by the jumping of the action potential among the unmyelinated spots along the axon. Because less current leaks out of the myelin-coated section of the membrane, the voltage gradient remains fairly high along the fiber; the membrane potential of the adjacent node is raised to threshold faster and initiates an action potential faster than in unmyelinated fibers. As a result, propagation velocity is higher, starting at 5 meters per second for myelinated fibers of 1 micrometer size and increasing directly with thickness to almost 170 meters per second for axons of 20 micrometers.[8] However, the addition

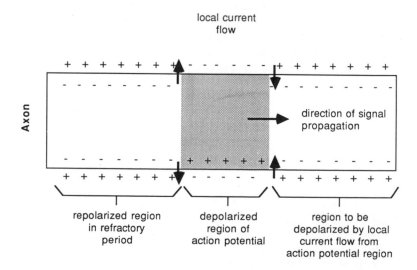

**Figure 5.2** Continuous propagation of action potential along an axon. (After Vander *et al.*, (1985), p. 205.)

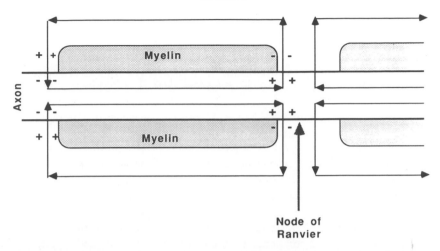

Node of
Ranvier

**Figure 5.3** Current flow in discrete signal propagation along a myelinated axon.

of a myelin coating increases the complexity of the system, since it is formed by cells specialized for this purpose. For this reason, the unmyelinated axon predominates in regions such as the cerebral cortex, where maximum conduction velocity is not critical, or where signalling occurs over relatively short distances.[9]

Our discussion of continuity versus discreteness in communication systems can be carried over from the neurological level to the global. The digitization of signals in communications networks—making them discrete rather than continuous—has resulted in improvements in the way data is transmitted. Conventional telephone lines use analog signals in the form of voltages modulated along a pair of wires. These analog signals are susceptible to degradation by noise and attenuation over long distances. When weak signals are bolstered by amplifiers along the telephone line, the noise is augmented along with the signal. In addition, the inertial effects of capacitance constrain the frequency of signal modulation, and thereby limit the bandwidth for information transmission.

Such analog facilities are being displaced by digital networks which offer more flexibility and resistance to degradation. When discrete voltage levels are used, it is simpler to deduce the original message even after it has been degraded through noise and attenuation. Hence, signals can usually be restored to their original crispness at each repeating station along a long cable.

The prevalence of digital communications is fueled by several other factors. First is the changing nature of communications: digitized data rather than voice signals now account for an increasing share of network traffic. Second, the increasing complexity of telecommunications networks has led to the use of computers to manage the networks. Computers allow for enhanced capacity utilization, as illustrated by the multiplexing procedure in which two or more incoming data streams are interwoven into a single outgoing stream. Another advantage lies in system robustness: data packets can be rerouted dynamically to avoid heavily-loaded links or inoperative nodes.

## Organization

The modules of an intelligent system must be organized in some fashion to achieve
coherent behavior. The most common structures are networks, hierarchies, and
layers. As we will see, hierarchies are a special class of networks.

### Network

A *network* is composed of a set of objects or *nodes*, and a set of *links* associating
pairs of nodes. Figure 5.4 shows an example of a network: each node might
represent a town, with the links depicting roadways among the towns.

Sometimes the links represent unilateral rather than bilateral relationships
among the nodes. In this case each link is represented by an arrow. To illustrate,
Figure 5.5 represents the transportation requirements of a manufacturing company.
Two plants manufacture components and feed them to an assembly plant. The
finished products are conveyed to two warehouses that supply a total of four dis-
tributors or wholesalers.

In general, a network represents the conceptual rather than physical relation-

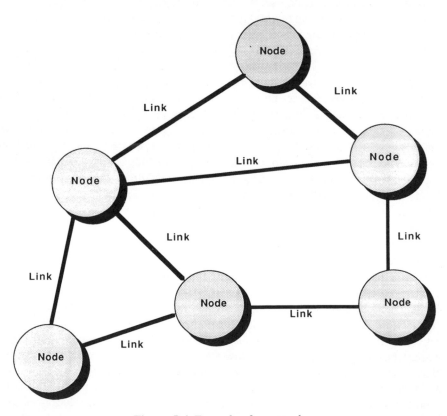

**Figure 5.4** Example of a network.

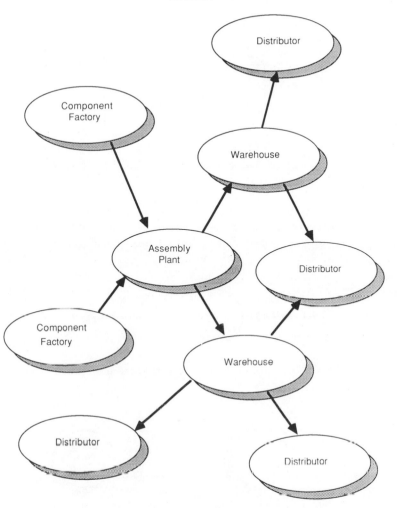

**Figure 5.5** Example of a network with directed links.

ships among objects. The distinction is illustrated in Figure 5.6. The first diagram depicts the spatial configuration of three objects: a chest contains a box which in turn contains a pyramid. The associated network is shown in the second diagram in the figure; each object is represented by a separate node and their conceptual dependencies by directed links.[10]

Networks are also often used to represent knowledge in natural language applications.[11] In this context, as with the previous examples, nodes represent objects while labeled links denote relationships among them. For example, the sentence "Fay gave the new red book to Jay" might be represented as shown in Figure 5.7. This sentence describes a specific event. To distinguish this activity from other events, it is given a unique serial number (say, 6) and thereby labeled *Event-6*. The event is an example (*Isa*) of a *Give* activity. The *Agent* is Fay and the *Beneficiary*,

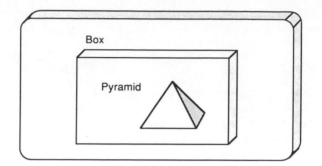

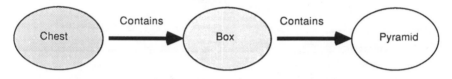

**Figure 5.6** Distinction between physical versus conceptual relationships.

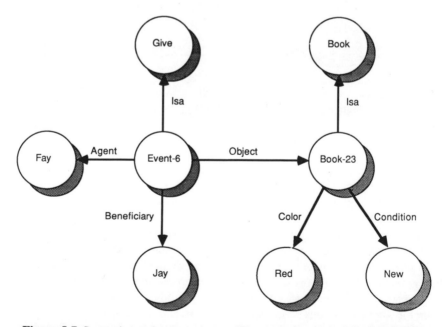

**Figure 5.7** Semantic net for the sentence: "Fay gave the new red book to Jay."

Jay. The *Object* of the event is a particular book (*Book-23*). This book is an instance of the type of object known as *Book*. Its *Color* is *Red* and its *Condition, New*.

## Hierarchy

A special case of a network is a *hierarchy*, where the nodes are grouped into a sequence of distinct bands or levels. In addition, communications are defined only among certain pairs of nodes in adjacent bands. One way to view a hierarchy is as a series of nested objects. According to this perspective, each node is regarded as a compound object consisting of subordinate nodes which in turn may be regarded as composite objects.

Hierarchies are found in all realms of natural and artificial systems. An example of a hierarchical pattern is the structure of the human nervous system. The major divisions are the central and peripheral systems, each heading its own hierarchy of subsystems (see Fig. 5.8).

The central network consists of the brain for sensory monitoring of the environment as well as high-level decision making. The spinal cord, in contrast, serves as a distributed center of control. It coordinates diverse tasks, such as mediating sensory input with motor output in reflex action. The spinal cord also serves as a conduit to relay signals to and from the brain.

The peripheral nervous system consists of afferent nerves which convey sensory input to the central system, and efferent neurons to transmit commands to bodily

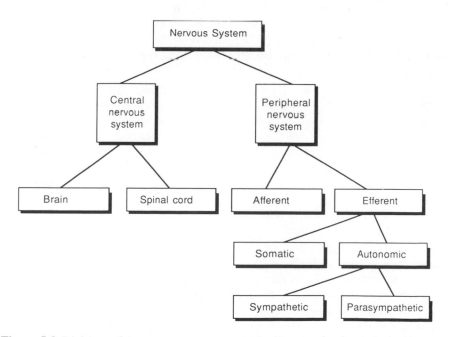

**Figure 5.8** Divisions of the nervous system. Each "leaf" or low-level node in the hierarchy can be partitioned further into specialized components.

parts. Efferent signals can stimulate the somatic neurons which innervate skeletal muscles. The efferent impulses also regulate autonomic organs such as the heart, which tend to operate without voluntary thought.

The autonomic division, in turn, consists of sympathetic and parasympathetic fibers. The former category consists of neurons that leave the central nervous system from the middle regions of the spinal cord. This group of fibers usually accommodates stressful situations such as environmental conditions requiring "fight or flight" responses. For example, the sympathetic division increases the rate of energy production and diverts blood flow from digestive functions to support muscle activity. In contrast, the parasympathetic fibers emanate from the central nervous system at the anterior and posterior ends of the spinal column. The parasympathetic division focuses on routine maintenance of bodily processes, such as digestion and urination.

Since periods of excitement and quiescence are mutually exclusive, the sympathetic and parasympathetic divisions operate reciprocally: the activation of one depresses the work of the other. For instance, sympathetic fibers increase heart rate while parasympathetic neurons reduce it; but at each instant only one type of fiber dominates the process.

Another example of a distinctive hierarchy is the visual system, which supports a corresponding information processing scheme. The analysis of information begins with the processing of electrical signals at the lowest level, and continues through numerous stages until finally a complex image emerges at the highest level. The system is based on the secondary sensory nuclei that transmit the raw data. The data consists of electrical impulses displaced from the resting potential as the result of a triggering stimulus. A preliminary version of the incoming information is composed at a higher level in the thalamus, perhaps in neurons located in the spinal gray cord. These preliminary images may be little more than approximate localizations of objects in space.

Enhancement of the data occurs in the mesencephalon where larger and more detailed pictures result; here, neurons respond only when several specific features are presented in tandem. At the highest level, a clear image is synthesized in the central region of the brain containing the gray matter. Through this type of progressive analysis we can recognize a hammer in a box full of tools, or the face of a friend. Information processing thus moves up different levels in a hierarchy, and in the process data become refined to represent increasingly complex combinations of features.

In the realm of behavior, goals and functions may often be decomposed into hierarchical structures. In this way, the overall purpose of a system may involve a set of supporting subgoals; for example, the fleeing and food-seeking behaviors of a salmon support its goal of survival.

A more involved example relates to the instance of a mobile robot.[12] The high-level functional requirement of the robot is to patrol a given territory in order to detect intruders and other exceptional conditions such as smoke. This function might be partitioned into three lower-level functional requirements as follows (see Figure 5.9):

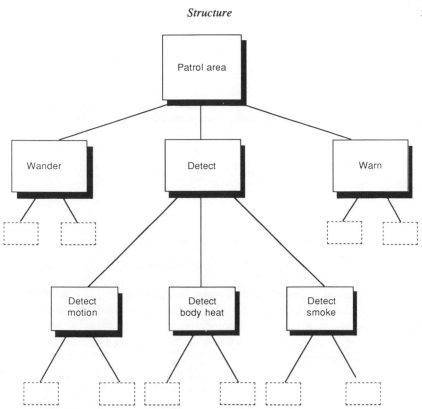

**Figure 5.9** Partial tree of functional requirements for a sentry robot.

- Wander through the security area.
- Detect intruders or other exceptional conditions.
- Warn a central station of such exceptions.

Each of these functional requirements may, in turn, be further decomposed. For example, the second low-level requirement may be partitioned into these activities:

- Detect motion.
- Detect body heat.
- Detect smoke.

The physical structure of the mobile robot must be designed to implement its functional objective. The hardware for the engineered system also has its own hierarchy, as illustrated for the sentry robot in Figure 5.10.

In general, the design process itself may be viewed in terms of a hierarchical structure. The design activity ranges from strategic concerns at the higher levels of decision making, to tactical concerns at the lower levels.

Consider the goal of building a general-purpose household robot. The overall strategy may consist of the construction of a series of special-purpose devices:

*Factors*

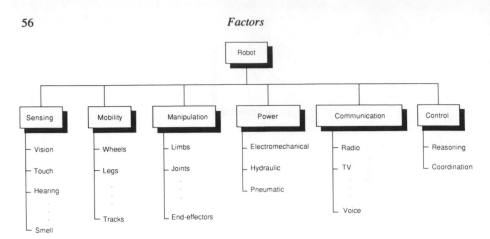

**Figure 5.10** Tree of physical components for a sentry robot.

robots to fetch objects, dust furniture, make beds, and so on. The development of each of these specialized devices may in turn be decomposed into a sequence of prototypes, progressing from the simplistic to the sophisticated. Every prototype will also require tactical decisions relating to the design of the sensors, manipulators, processors, and myriad others.

The parallels are obvious in many other domains. In the realm of industrial decision making, suppose the objective of a corporation is to become a world leader in medical instrumentation. Its short-term strategy for obtaining the technological capability may be through licensing and acquisitions, and long-term capability through a series of internal think tanks. Its marketing strategy may take the form of testing the field in several small countries in Africa or Asia; then applying the lessons learned to larger markets such as the United States. Each of these strategic moves will entail tactical considerations, such as the selection and timing of products to be introduced, or the portfolio of publications to use for display advertising.

## Layer

In a layered organization, the system components belong to different levels. Components on adjacent levels can communicate through interfaces. The interfaces mediate the communication among layers, as well as hide the structures of the individual layers from each other. In layered systems, higher-level procedures build on lower-level ones.

The human brain is a layered system, and each of its three separate levels corresponds to a different era in the brain's evolutionary history: reptilian, limbic, and neocortical. The reptilian, or R-complex, component is located centrally and is prominent even among mammals. It controls basic functions such as respiration and some aspects of temperature regulation. It also influences aggressive behavior, territoriality, ritual, and establishment of social hierarchies. The limbic system—containing the thalamus, hypothalamus, and pituitary—adjoins the R-complex. This layer serves as a center for emotional responses, as well as memory storage and recall. The neocortex in turn surrounds the limbic system and is associated with higher-level functions such as reasoning. A schematic of the layered structure is shown in Figure 5.11.

An example illustrative of these functions is the problem of finding a path leading out of a forest. The high-level strategy, formulated in the neocortex, might consist of a direction (e.g., western heading), and negotiation of known obstacles (crossing a river). The limbic level is responsible for memories of old experiences, rules of thumb, and perhaps fear of predators. Meanwhile the R-complex level deals with basic functions such as respiration, perspiration, and coordination of bodily movement.

The parts of the brain used in language production and understanding are also layered, interacting with each other in a similar fashion. The basics of utterances originate from the region of the brain known as Wernicke's area. From there the bundle of nerves known as the *arcuate fasciculus* transmits these messages to Broca's area, a neighboring region of the brain. Broca's area programs the actual sounds and sends messages to the area of the cerebral cortex, which controls the face. The cortex then activates the appropriate muscles to produce speech.

The human retina is organized into several horizontal layers, as well. These layers are made up of neurons and photoreceptor cells, which are distinguished both physiologically and functionally. The most "superficial" layer of receptors contains the outer segments of rods and cones, which respond to incoming light. Rods are more sensitive to light than cones but less sensitive to color, and possess limited

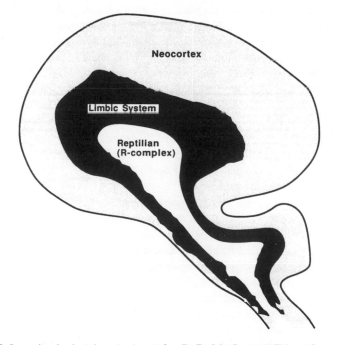

**Figure 5.11** Layering in the triune brain. (After P. D. MacLean, *A Triune Concept of Brain and Behavior,* University of Toronto Press, Toronto, 1973.)

powers of resolution. Cones have higher thresholds for response to light stimuli and
are geared toward visual acuity and color vision in good light. The rods and cones
interface with bipolar neurons extending in two directions: one leading to the
periphery, and one to the brain. The articulation between the two types of cells is
enhanced and integrated by "horizontal" cells.

The bipolar cells conduct electrical impulses to the retinal ganglion cells; the
axons of this latter group transmit the impulses along the anterior surface of the
retina to the optic nerve. The interface between bipolar cell and retinal ganglion cell
is aided by *amacrine cells*, which, like the horizontal cells, perform integrative
functions and do not possess axons.

The different levels of computer languages provide an artificial example of
layering (see Figure 5.12). The bottom-most layer, upon which all the languages
rest, is the hardware of the machinery itself. The progression of computer languages
is defined by a machine language specific to a particular model of hardware. This is
followed by an assembly language having straightforward ties to the machine language, but eases the burden of programming for a human user. Atop this layer sits a
procedural language such as Pascal, C, or Prolog, whose advantage lies in its
portability or hardware independence. Finally, the procedural language may be used
to develop a nonprocedural language specific to a given application. Examples in
this category are software for financial spreadsheets or packages for creating draw-
ings.

These languages are used to translate problems from the level of natural language to one that computers can understand. The sequence of languages represents
a total ordering in which each language represents a different layer. The different
layers are basically isolated from each other. Within the procedural language layer,
for instance, the statements of the programming language interact only with each
other. Their communication with the assembly language layer is mediated by the
procedural-assembly interface, or compiler.

## Hierarchy versus Layer

The structure of a hierarchy is characterized by its span of control and the number of
levels. The span for a particular node is specified by the number of nodes directly
subordinate to it. For a given number of nodes, the span is obviously related to the
number of levels in the hierarchy.

For example, consider a manufacturing plant in which the central computer
coordinates 5 departmental computers that collectively monitor 30 machine tools. In
this case the span for the central unit is 5, while the average span for each departmental controller is 6 and the number of levels is 3. If the 30 machine tools are to be
monitored directly by the central computer, its span is significantly higher and the
number of levels drops to 2.

The appropriate span is determined in part by the relative complexity of interactions at adjacent levels. For example, a supervisory unit that spends a large amount
of time monitoring subordinate nodes can oversee fewer units than one that monitors relatively autonomous units.

Another determinant of the span relates to flexibility. A system whose internal
organization changes frequently may be better served by a low average span. The

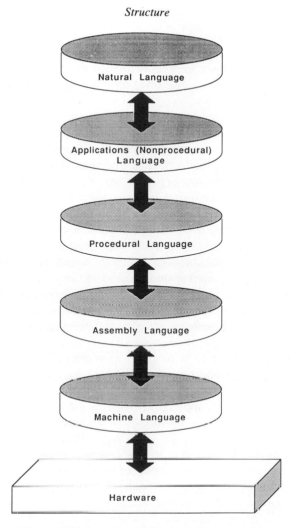

**Figure 5.12** Layering of programming languages.

larger number of "supervisory" nodes will increase the likelihood that the subsystem that needs to be changed will correspond precisely to the jurisdiction of a particular node. For example, a military organization for which the span of control is 10 at each supervisory node can be more readily deployed for engagements of varying size than an army in which each platoon is made up of 1000 soldiers.

In a similar way, the number of levels in a layered organization depends on the degree of complexity that can be handled at each layer as well as the requisite degree of flexibility. Increasing the number of layers simplifies the structure of each layer and facilitates modifications. But the advantages are gained at the cost of inefficiencies in interlayer communications.

For example, the Open Systems Interconnection Reference Model developed by the International Standards Organization specifies a standard for telecommunications protocols. The model consists of 7 layers, ranging from the Physical Layer for interpreting electronic signals, up to the Application Layer for domain-specific software. The large number of layers accommodates the adoption of the Reference Model to specific software protocols. However, actual implementations may combine two or more layers into a single level for efficiency in system operations.

*Database Architectures.* Both hierarchies and networks have been used as the conceptual foundation for database management systems. These database systems build on an architectural scheme for organizing voluminous amounts of data. Networks are more flexible than hierarchies in that links may be forged between any two nodes. To illustrate, a network may have one or more "superior" nodes, while a hierarchy can have at most one. For this reason, the hierarchical systems developed in the 1960's began to give way to the more flexible network models in the 1970's.

The disadvantage of these database architectures lies in their inflexibility. Once a database is established according to a particular hierarchical or network format, it is cumbersome—if not impossible—to modify or extend the structure.

These have both yielded market share in the 1980's to a newer relational model based on tabular representations of data. Having no predefined relationships among different "relations" or modules, the flexibility and conceptual simplicity of this approach has earned it a prominent place in the theory and practice of database systems.

These different approaches to database management come in three primary flavors defined by their perspective of the world: hierarchical, network, and relational schemas.

The *hierarchical* model is appropriate when the underlying structure exhibits a 1-to-$N$ relationship in which one superordinate node links with an arbitrary number $N$ of subordinate elements. The value of $N$ may vary from zero to infinity. Conversely, each node or entry has at most one superior—except the topmost or *root* node, which has none. Each node consists of one or more records, which may be viewed as "cards" containing data in a predetermined format.

A sample database for the Millennium Corporation is shown in Figure 5.13 At the first level below the root node, there are two record types: **Product Groups** and **Customer Types.** The first record type contains three entries: **Nutrition, Apparel** and **Vehicles.** Each of these, in turn, refers to subordinate record types containing more entries. In a similar way, the **Customer Types** node has two entries which, in turn, point to offspring nodes.

The hierarchical database schema is simple to implement and convenient for responding to queries such as "What vehicles does the Corporation produce?" On the other hand, the hierarchical schema is inconvenient for responding to questions such as "Which customers ordered space yachts in the last 6 months?" and "What products did Al Andromeda buy last year?" To accommodate the former question, each record in the **Vehicle Products** record type would have subordinate nodes for each customer. To answer the latter question, each entry in the **Customers** stack would have subordinate nodes for each product. The redundancy and computational inefficiency of such a strategy is obvious.

These difficulties are addressed by the *network* database schema, which is appropriate for representing $M$-to-$N$ relationships. In other words, each node may have more than one superior node as well as any number of subordinate nodes.

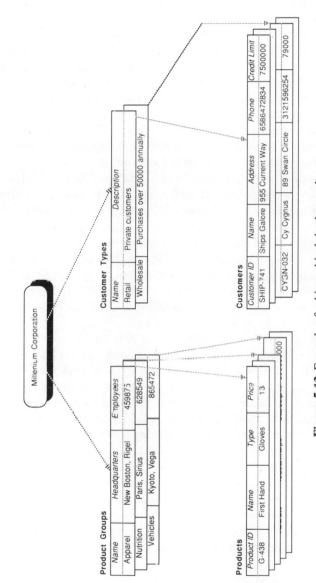

**Figure 5.13** Example of a hierarchical database schema.

Figure 5.14 illustrates how the network schema can accommodate such a relationship. This is done by defining an **Orders** record type that links both the **Products** and **Customers** nodes. For each **Products** record, it is possible to travel down to the corresponding **Orders** record, and from there move up to the referent **Customers** record; and vice versa. Since a customer may have purchased more than one item on a single order, each order may point to more than one product record.

Despite the increased power of the network model over the hierarchical schema, it is still short on flexibility and ease of use. What happens, for example, when a motorized flotation device is to be moved from the apparel product line to the Vehicle group? Are all the relevant pointers in the **Orders** record type to be changed? Perhaps a more chronic difficulty comes from the fact that (a) searches through the database can be efficient or not depending on the fit between database structure and the user queries, and (b) the programmer and/or database administrator must take explicit account of the structure of the database in order to traverse through the network and obtain the desired answer.

These difficulties are eliminated in the *relational* schema, which views a database as a set of independent record types. Hence, any record type may be added, modified, or deleted at will, with little regard for the structure of the other record types.

This approach uses no explicit pointers to link records to each other. Instead, each record has a unique identifier or *key* that is used to access the record when needed. For example, the first **Orders** record shown in Figure 5.15 has 50801 as its key. The customer identification (ID) for this order is ANDR-493. If, for example, the address of this customer is desired, then the relevant record in the **Customers** record set is sought using ANDR-493 as the key.

The database schema example illustrates not only the issue of centralization, but also of networks and layers. The hierarchical database schema may be viewed as a special case of the network model in which only 1: $N$ relationships are allowed rather than $M : N$. In turn, the relational model may be viewed as a degenerate hierarchical schema with only one level and therefore no internode linkages.

The relational model provides for no pointers or markers between pairs of records. Hence each record must be sought using some search technique: for example, the **Customers** record type must be searched using the ANDR-493 key. The inefficiency of such an approach is apparent.

As a result, the underlying software for a relational database management system may utilize a network approach, with full use of pointers. This implementation level is kept hidden from the user's view. In this way, the system utilizes a layered approach: the conceptual model of the database is a relational scheme, but the implementational model is that of a network.

The structure of a system accommodates the processes that implement the objectives of the system. This attribute is closely affiliated with, but separable from, processing issues such as the patterns of control. For example, one company may be formally organized into a hierarchy with tight control from headquarters; another hierarchical organization in the same industry may be patterned as a loose confederation of largely autonomous divisions.

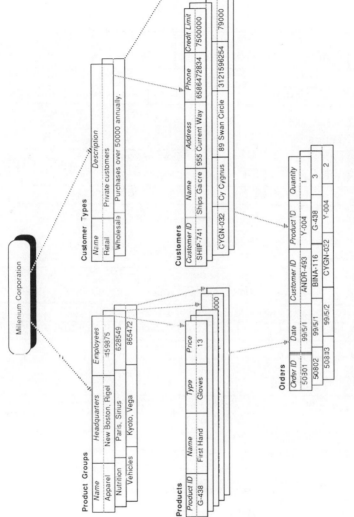

**Figure 5.14** Example of a network database schema.

**Product Groups**

| Group ID | Name | Headquarters | Employees |
|---|---|---|---|
| APP | Apparel | New Boston, Rigel | 459875 |
| NUT | Nutrition | Paris, Sirius | 628549 |
| VEH | Vehicles | Kyoto, Vega | 865472 |

**Products**

| Product | Group ID | Name | Type | Price |
|---|---|---|---|---|
| G-438 | APP | First Hand | Gloves | 13 |
| S-972 | VEH | Trans Warp | Starship | 890000000 |
| Y-004 | VEH | Light Rider | Space Yacht | 2900000 |

**Customer Types**

| Type ID | Name | Description |
|---|---|---|
| RE | Retail | Private customers |
| WH | Wholesale | Purchases over 50000 annually. |

**Customers**

| Customer ID | Name | Address | Phone | Credit Limit |
|---|---|---|---|---|
| ANDR-493 | Al Andromeda | 47 Galactic Lane | 3195472316 | 49000 |
| BINA-116 | Bo Binary | 10 Stellar Drive | 0110011100 | 37000 |
| CYGN-032 | Cy Cygnus | 89 Swan Circle | 3121596254 | 79000 |

**Orders**

| Order ID | Date | Customer ID | Product ID | Quantity |
|---|---|---|---|---|
| 50801 | 99/5/1 | ANDR-493 | Y-004 | 1 |
| 50802 | 99/5/1 | BINA-116 | G-438 | 3 |
| 50803 | 99/5/2 | CYGN-032 | Y-004 | 2 |

**Figure 5.15** Example of a relational database schema.

## Parallelism

*Parallelism* refers to the degree of simultaneity in form or function. For example, a set of struts may support a building in parallel, thereby dividing the load among them. In contrast, a set of struts stacked vertically atop each other would represent a sequential configuration. In the second case, each strut takes full burden for the load.

In a similar way, information processing can occur either in sequential order, as a series of tasks, or in parallel, where many related processing functions occur simultaneously, but via different pathways. The configuration of the system, whether sequential or parallel, depends on the desired function.

The spinal cord, for example, is organized into parallel pathways so that information reaches its destination as quickly as possible. The axons of neurons run along the length of the spinal tracts in two directions. They either descend in order to transmit information from the brain to the spinal cord, or ascend to transport information in the opposite direction.

Moreover, each pathway contains fibers that perform the same function; the information from light tactile receptors in the skin proceeds up its own route, for example. The partitioning of fibers into separate bundles speeds up the transmission and processing of information. Such channels enable the signals to travel directly to the place where they will be processed, thereby minimizing the effort necessary to integrate data.[13] In addition, this scheme ensures that all the relevant signals reach the processing center simultaneously.

Every information network possesses advantages and limitations. In higher animals and humans, a comparison of the endocrine and nervous systems provides an excellent example of the trade-offs involved. All information in the endocrine network is disseminated throughout the system by the bloodstream, even though only the target cells can decode the message. Further, the endocrine system operates under the limiting constraint of the speed at which blood travels through the body. As a result, the endocrine system is much slower and less efficient than the nervous system in sending messages to specific locations. Yet the endocrine network can broadcast information widely throughout the organism without requiring the establishment of a separate connection to each target center. For example, the growth hormone acts upon cells at many different locations simultaneously, as do the sex hormones known as gonadotropins.

In contrast, neurons deliver their signals only to very specific locales. To their credit, they accomplish this much more quickly and accurately than the endocrine system. In addition, the structure of the neural components allows for parsimony in the generation and use of neurotransmitter substances.

Each of the parallel systems is particularly well-suited toward a different information-transfer task. The nervous system is the network of choice for fast and accurate signalling to a small area, while the endocrine system is better adapted for information that must be transmitted over larger areas and for longer periods of time.[14]

Some biological processes are better served by a sequentially ordered string of events. In the case of the knee jerk, each sensory neuron from the muscle spindle receptor senses changes in muscle length and tension, then synapses with motor neurons that stimulate the contraction of the muscle. The pathway for this quadriceps reflex consists of neuronal connections organized in a precise sequence.

The anatomy of the reflex arc is as follows:

- A sensory receptor in the muscle spindle responds to environmental stimuli, such as a tap on the quadriceps tendon.

- The afferent neuron transmits electrical signals to the central area of gray matter in the spinal cord.
- Inside the spinal cord the afferent neuron synapses with the lower motor neuron.
- The motor neuron transmits an efferent impulse back to the muscle fiber and initiates contraction.

In the specific case of the quadriceps reflex, the tap on the tendon stretches the quadriceps muscle, activating the muscle spindles. The spindles initiate transmission of electrical signals through afferent nerve fibers. These fibers then stimulate the motor neurons. The eventual result is muscle contraction.

For the reflex process to function properly, these events must occur in the prescribed sequence so the appropriate neuronal connections can be made. Since the events do occur singly and in order, the reflex arc can be described as *serial processing*.[15]

A similar rigidity characterizes the process through which the application of a stimulus is transduced into neural code. An appropriate stimulus, applied to the nerve ending, triggers a local increase in sodium permeability. Due to this increase in membrane permeability, and to the concentration gradient created by the $Na^+/K^+$ pump, sodium ions stream into the neuron. This influx produces a local depolarization of the receptor's membrane. The size of the generator potential (i.e., change in voltage) is always smaller than that which characterizes the action potential, but varies with the intensity of the stimulus. Eventually, the generator potential reaches the threshold for initiating an action potential. At that point, an action potential is generated and conducted along the axon.

The events may be described as occurring in serial order, since each requires the occurrence of the preceding phenomenon. Without the application of the proper stimulus, no increase in the membrane's permeability to sodium occurs. Without the influx of sodium ions, no depolarization occurs. Hence, a sequential order of these occurrences must be established and maintained if an action potential is to be initiated.[16]

Conventional computers, too, are sequentially ordered. They are based on the so-called von Neumann architecture which regards computation as a series of steps proceeding one after another and operating on a stream of data from memory. The fastest machines today have transition times on the order of nanoseconds. However, the von Neumann architecture is not likely to yield much higher speeds, because of the fundamental constraints imposed by the physics of the hardware. One such limitation springs from the quantum mechanical properties that constrain the physical dimensions of minute devices. Another limit is the speed of light, a constraint especially restrictive in the transfers of data between the processor and memory.

The modern computer has enormous potential for information processing in comparison to the brain. The human brain contains approximately 10 billion neurons, each of which may switch at most a thousand times a second. As a result, the brain has the capacity to perform roughly 10 trillion operations per second. In contrast, a digital computer might possess a billion transistors, each of which might switch a billion times per second, for a total of one quintillion operations per second. By this argument, the computer should outperform the human brain by a factor of 100,000.

However, the computer is no match for a human or even a simple rodent in many tasks such as physical coordination or scene recognition. The deficiency in the computer lies in the lack of parallelism in its architecture. This results in the storage of data and programs in one section of the machine, and their processing in a separate section:

> The two-part architecture keeps the silicon devoted to processing wonderfully busy, but this is only 2 or 3 percent of the silicon area. The other 97 percent sits idle. At a million dollars per square meter for processed, packaged silicon, this is an expensive resource to waste. If we were to take another measure of cost in the computer, kilometers of wire, the results would be much the same: Most of the hardware is in memory, so most of the hardware is doing nothing most of the time.[17]

A solution to the "von Neumann bottleneck" of data transfer limitations is to employ multiple computing units in parallel. This philosophy is illustrated by vector processors, which can perform an identical sequence of operations on perhaps a dozen data streams. A more dramatic approach lies in the Connection Machine, a device that can harness the power of thousands of processors.[18] These units can execute similar instructions on different data units. An embryonic methodology is the integration of semi-independent processors that can all execute different instructions on distinct data. Such parallel architectures, however, suffer from the dearth of software techniques that can take full advantage of the hardware. To this end, existing languages such as the logical programming language of Prolog[19] are being extended for parallel interpretation in hardware.

## "Strategy" versus Structure

A common theme in system design is the interplay between strategy and structure. The term *strategy* is used in several senses. The two most common are as follows:

- *Purpose*. Strategy is used as a synonym for a goal or objective. An example is found in the expression, "Our corporate strategy is to become a world leader in consumer electronics." In this sense, strategy is a basic parameter of the system design.
- *Design*. A strategy is viewed as an overall approach to a design that fulfills some purpose. Examples are found in comments such as "The robot's strategy is to use a cart to move the heavy box," or "Our strategy for institutional excellence is to develop a sense of professionalism among all the employees." In this case, strategy is a metalevel concept relating to the process of design— the approach to implementing a goal for a particular system.

Hence, the concept of strategy can straddle the dimensions of intelligent system design, whether in terms of space, structure, or any of the other attributes discussed in Part II. Therefore, the phrase "strategy versus structure" denotes distinct concepts in different contexts.

In this section we take the view of a *strategy* as the set of high-level decisions that define a design or approach dedicated to some objective. This perspective is depicted in Figure 5.16.

According to this view, the interplay of strategy versus structure is a conceptual hybrid. On the one hand, strategic concerns are high-level design issues that may be decomposed into lower-level tactical decisions required to support their respective strategies. In other words, strategy is a metalevel issue relating to the process of designing a system—or, more generally, of developing an approach to fulfill some objective.

On the other hand, structure is a specific dimension of the design itself, as discussed previously. Hence, the juxtaposition of strategy and structure does not signify a trade-off between two aspects of a similar nature, as do "time and space," or "mechanism and process."

This interpretation differs from popular usage as illustrated in management literature. In this field, "strategy versus structure" is used to denote the interplay between the objectives of an organization and its internal structure.[20]

In organizational design, as in other fields, the structure of an institution is determined by its purpose. While this is true in the long run, it is also likely that short-term tactics are influenced by the organizational structure. In the 1970's, for

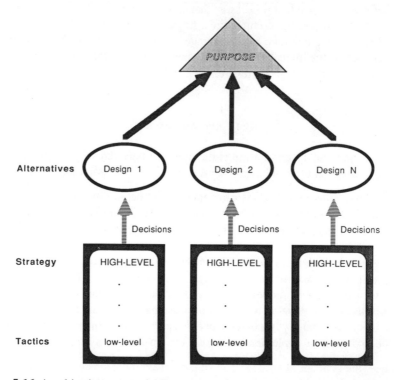

**Figure 5.16** An objective may be fulfilled by any of several alternative designs. Each design may be realized through a set of decisions ranging from high-level strategies to low-level tactics.

example, the Xerox Corporation failed to fully utilize artificial intelligence technology to develop smart copiers. This was due in part to its internal structure, which separated basic research from product development activities. The advanced computer science work was pursued at a research center in Palo Alto, California, and little attempt was made to incorporate the new technologies into products under development in Rochester, New York.

In conclusion, strategy is used in two senses, as a purpose or an approach to fulfilling some purpose. In the former sense, structure is determined by strategy; in the latter, structure is only one aspect of system design.

## Summary

The purpose of a system defines its structure. The architect must determine the most effective way to handle complexity by partitioning the system into modules, to increase performance through parallelism, and to merge low-level tactics with high-level strategies.

- Lower-level objectives may be determined in part by precedent, including previous system structure. An example is a manufacturer of motorcycles that parlays its engineering expertise to produce automobiles.
- Modularity refers to the chunking of a system into smaller components. The granularity of the components defines the smallest meaningful units.
- Coarse granularity allows for convenience of reasoning and expression while fine granularity permits precision or refinement in behavior.
- A continuous parameter such as a voltage signal may be viewed as a limiting case of a discrete attribute.
- The modules of a system must be organized in some way to fulfill high-level objectives. The common forms of organization are networks, hierarchies, and layers.
- Parallelism refers to the replication of a module, whether in form or function. Parallel structures are used to accelerate a process, as in computation, or to amplify a parameter, as for structural strength.
- Parallelism is appropriate when the need for intermodule communication is low.
- The term *strategy* is often used in two senses, referring either to the purpose of a system or to the approach to attaining some purpose. In the former sense, strategy as purpose is one dimension of the design; in the latter, strategy as approach is a metalevel description of the process for generating an appropriate design.
- Structure is defined by purpose or "strategy"; but the structure in turn may determine short-term tactics or "strategies."

# 6

# Time

*The deterministic laws of physics, which were at one point the only acceptable laws, today seem like gross simplifications, nearly a caricature of evolution. . . . The future is not included in the past. Even in physics, as in sociology, only various possible "scenarios" can be predicted.*[1]

Ilya Prigogine

*Psycho-history dealt not with man, but with man-masses. It was the science of mobs; mobs in their billions. It could forecast reactions to stimuli with something of the accuracy that a lesser science could bring to the forecast of a rebound of a billiard ball. The reaction of one man could be forecast by no known mathematics; the reaction of a billion is something else again.*[2]

Isaac Asimov

Any attempt to design an intelligent system for information processing, whether natural or artificial, requires a consideration of a number of referent attributes. The issues to be considered are generic, such as the need to respond in timely fashion to environmental stimuli. However, for a given parameter such as timeliness, two equally effective agents may exhibit diverging characteristics, perhaps one reacting quickly and another only slowly.

The attributes of intelligent systems often oppose each other, in the sense that inclusion of one will necessitate the exclusion of another. In such cases, a trade-off between two conflicting attributes occurs; the system architect must then decide which attribute is more desirable in terms of achieving the stated goals of the system, subject to any other constraints. Thus, the central issue in the design of an information-processing system often boils down to finding the optimal balance within pairs of contrasting extremes.

## Static versus Dynamic

A static condition is one that remains constant over time, while a dynamic one does not. Any intelligent system of reasonable complexity will likely exhibit features of both types.

The relationship between static vs. dynamic characteristics is a common theme in intelligent action. For example, a robot may be able to lift heavy loads when it stays rooted to one spot, but is incapable of transporting such loads for any appreciable distance.

On the other hand, certain systems are effective dynamically but not statically. A bicycle and its rider are stable in motion but not while at rest. Sometimes it is helpful to view the static situation as a limiting case of a dynamic one, in which the period of oscillation approaches infinity. This is the view taken at times by control systems engineers when regarding constant loads on a system as opposed to cyclic ones. In a similar way, the constant voltage provided by a conventional battery may be regarded as a special case of an alternating voltage source.

The contrast between static and dynamic aspects also appears in declarative versus procedural interpretations in programming languages. Statements in a computer language may be viewed in both declarative and procedural terms. This is especially true of Prolog, a language for artificial intelligence based on mathematical logic.[3] This language uses three types of statements: assertions, rules, and queries.

An assertion is a declarative statement that is taken to be true. Examples of assertions are:

**fruit(orange).**
**flat(earth).**
**parent(alex, brenda).**

The first statement encodes the fact that that an orange is a fruit. The second item asserts that the earth is flat; this example highlights the fact that Prolog will assume as valid any statement that is asserted, regardless of its validity in the external world. The third item above may be used to declare that Alex is a parent of Brenda.

A rule is a statement in Prolog that relates one item to other items through a conditional implication. The rule

**edible(X) if fruit(X).**

relates fruitage to edibility. A declarative reading is "For any object **X**, **X** is edible if it is a fruit." The rule may also be given a procedural interpretation: "To show that some object **X** is edible, show that it is a fruit."

The conditional part of a Prolog rule may be a compound structure consisting of two or more predicates. The rule

**feasible(System) if**
    **satisfies(System, Requirements) and**
    **satisfies(System, Constraints).**

might be used to encode the knowledge that a feasible system is one that satisfies its functional requirements and constraints.

The third type of statement, the query, can also be given a declarative or procedural interpretation. A query is initiated in response to the prompt "?-" from Prolog. The inquiry

**?- edible(orange).**

may be viewed in declarative or passive terms as "Is an orange edible?" A procedural interpretation is, "Show, if possible, that an orange is edible." If this item of knowledge exists in the database of facts, or is derivable through other rules, then Prolog will respond with a "yes"; otherwise the response is "no." If the knowledge base happens to contain the preceding rules and facts, Prolog will determine that an orange is a fruit and is therefore edible.

## Entropy

Chaos refers to the degree of randomness or disorder in a system. Physical mechanisms tend to become more disorderly over time, a notion that is embodied in the concept of entropy. The cause of the entropic tendency may be traced to dissipative mechanisms such as friction. An intelligent system, to exhibit purposive behavior other than deliberate randomness, must embody structure. In so doing, it must constantly combat a universal tendency toward disorder. The order in an intelligent system may be characterized by a number of attributes such as organization, adaptation, and self-repair.

*Organization* may be induced internally or externally. When structure or pattern arises as a result of an internal propensity, we call it *self-organization*. This characteristic is prevalent in biological systems; consequently *self-organization* is the term often used to characterize life itself. Structure is found at all levels of biological entities, from organic compounds to organ systems. The Austrian physicist Erwin Schrödinger has argued that "the device by which an organism maintains itself stationary at a fairly high level of orderliness . . . really consists in continually sucking orderliness from its environment."[4]

Examples of this internal organization may be found throughout the human body. On a microbiological level, the cells constituting the body components possess highly ordered substructures, such as mitochondria; on a more macroscopic level, the layout of the various organ systems is also characterized by a high degree of organization.

Externally induced organization can be attributed to some foreign source. For example, a manufactured item, once fabricated, embodies characteristically ordered patterns that decay over time. A watch runs down periodically and must be re-energized; eventually (and often too quickly) it breaks down.

The goal, and perhaps the *raison d'être*, of most biological systems seems to be the maintenance of constant composition of the system's internal environment in the face of changing external conditions. Hence *adaptation* would be a key attribute for any biological system. Yet order is also a prerequisite for such flexibility. When in a disordered state, the system cannot even begin to receive information about changing external conditions, much less process and respond to the input. Adaptability is a feature of integrated systems functioning in an organized fashion.

The feature of *self-repair*, which most biological systems possess, also relates to the tendency toward order. Within the human body, the mechanisms for blood clotting and bone fracture repair appear to be examples of this attribute. Both of these procedures consist of an orderly sequence of events, leading from initial injury to complete healing.

Disorder may be discussed in terms of two subissues: natural entropy and deliberate entropy. *Natural entropy* refers to the tendency of all systems, according to the second law of thermodynamics, toward increasing disorder or randomness. The net effect of such a tendency is to reduce useful energy. Yet biological systems, like certain inanimate structures such as stellar systems, defy this rule. They often possess a great deal of organization and have not only survived but have managed to prosper in an often harsh and violent universe.

*Deliberate entropy* refers to the instances in which disorder is desirable. An example of such an instance would be the shuffling of a deck of cards to introduce randomness. Another case is the deliberate scrambling of communication signals between sender and recipient to foil the interception and interpretation of the message by some unauthorized third party.

Randomness may be viewed as an extreme case in a spectrum of structures. The constancy of the lunar surface over the eons embodies a high degree of stasis. At the other extreme is the fiery violence of sunspots and stellar catastrophes. Most activities in our everyday experience exhibit a moderate level of structure, ranging from language and music to morality and law. These must exhibit some pattern to serve their purpose, yet excessive structure can hamper the very goals they are intended to serve. For example, laws must be comprehensive and precise, yet they must leave some room for interpretation; otherwise justice would be blind to unforeseen circumstances and the changing social environment. In the musical realm, a tune must exhibit some structure; otherwise it is heard only as noise. On the other hand, a tune that is too orderly becomes predictable and boring.

Many pheonomena exhibit graduated levels of randomness. An example of this is the laminar flow of fluid: water flowing gently in a stream can soon begin to swirl in vortices, which may crash then may crash against boulders, breaking up into spray and mist.[5] The interpretation of determinism and nonpredictability as special cases of a spectrum of order is depicted in Figure 6.1.

Order can arise spontaneously in natural phenomena, as illustrated by chemical reactions that involve the genesis of structure from previously homogeneous materials. One such instance lies in the Belousov-Zhabotinskii chemical reaction that occurs at room temperature following the mixture of three reagents: potassium bromate; malonic or bromomalonic acid; and ceric sulfate or a similar compound dissolved in citric acid. The ensuing reaction leads to wave-like activity and the creation of highly symmetric patterns on a formerly homogeneous surface. Moreover, the reaction oscillates: the starting materials yield the products of the reaction, which in turn reproduce the original reagents. Hence, order is generated over both space and time by this process.[6]

Oscillatory reactions also occur in biological systems, as exemplified by regulatory processes within cells. The catalytic processes for protein fabrication are swift when compared to the inherent stability of the resulting proteins. Consequently, the amount of protein in a cell may be excessive, thereby activating other mechanisms for arresting the rate of protein formation. The negative feedback can result in oscillations in the level of protein.[7]

How do biological agents and certain inanimate systems become more orderly over time? One way or another, order must be paid for. According to the second law of thermodynamics, an isolated system—one that does not transport materials or

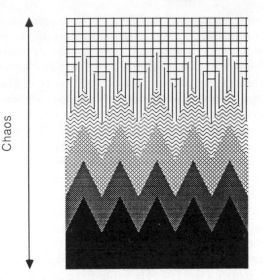

Chaos

Order;
Predictability;
Determinism

Disorder;
Randomness;
Nondeterminism

**Figure 6.1** Randomness and predictability as endpoints on a continuum of haphazard phenomena.

energy across its boundary—must tend toward increasing entropy or at best remain the same. Any system together with its universe constitute an isolated super–system; so the increase in order within a system must occur at the expense of increased disorganization in some other part of the system or environment.

In the case of the Belousov-Zhabotinskii reaction described above, some waste products are formed in each cycle, and the active ingredients are eventually depleted. One way to view the phenomenon is to regard the initial order in the separate groupings of the reagents, plus their respective internal energies, as being transformed into the kinetic energy of movement. This energy eventually turns into frictional losses and dissipates into the environment as heat.

In the domain of biology, catalytic energy is derived from ingested food whose useful energy was obtained in the first place from solar radiation. But the sun itself is engaged in an irreversible process of aging, and will eventually fade over the course of billions of years. The degeneration of the stellar battery implies a dim outlook for intelligence and life itself over the long term.

In the interim, however, there is much we can do to arrest the tide of disorder in our local environment. An appropriate strategy in designing an intelligent system lies in generating one that works with, instead of against, the tendency toward increasing disorder. In synthesizing any system, the designer must select the appropriate modes of organization as well as the optimal degree of structure for each mode.

The tendency toward decay seems to characterize not only mechanical systems, but large-scale organizations as well. Economic activities involve transformations that consume structure and create disorder. For example, a block of pure iron has greater structure than the original iron ore; but the extractive process consumes useful energy in a transformation whose net effect is to increase entropy.

Hence any economic system tends to decay over time unless rejuvenated through new inputs such as physical resources or technological capabilities. The management philosopher Peter Ferdinand Drucker has defined the essence of business activity as the process of counteracting the diminishing productivity of capital. In his view, the only way to protect an economy from the ravages of entropy is to keep renewing the productivity of capital by shifting resources to more productive uses.[8] For example, mining equipment might be employed to extract metals from a rich lode. But the high concentration of metal ore will eventually give way to regions of low concentration, and the mining facilities will be better utilized in another location where the ore is more concentrated or structured.

The English historian Arnold Joseph Toynbee has studied the 26 civilizations known to us, including the current Western culture. His work examined the pernicious effects of entropy at the social level. The growth phase of a civilization is characterized by diversity and the vitality it engenders; in this developmental stage the society is led by a creative minority within the population. Over the course of time, however, the creative minority establishes itself as the ruling segment of the society and eventually succumbs to ossification:

> The disintegrations of civilizations . . . descry a tendency toward standardization and uniformity: a tendency which is the correlative and opposite of the tendency toward differentiation and diversity which we have found to be the mark of the growth stage of civilizations.[9]

The analogy between the societal and physical realms is apparent. In due course, molecular activity in an isolated system becomes random, corresponding to the equilibration of temperature within the system. Although the total energy of the system remains unchanged, the uniformity of temperature implies that no useful energy is available.

Hence, in the realm of civilizations, the failure of the ruling minority to remain creative and dynamic leads to social decay and the ultimate disintegration of social unity.[10] It seems ironic that standardization and homogeneity—which we often regard as epitomes of structure—lead to disruption and decay in the societal sphere.

The preceding example highlights the fact that many dimensions are usually relevant in considering the design of an intelligent system. For the social sphere in particular, what are the design implications for the tendency of civilizations to become homogeneous and anemic? One strategy would seem to lie in the emplacement of mechanisms to ensure continuing diversity and vitality, such as the induction of new members into the ruling minority from the larger pool of the indigenous majority or even from foreign cultures. Surely the constant flux of immigration and the resulting cultural diversity has been an important factor in the vitality of the United States during its first two centuries.

From this perspective, the societal example highlights the need to maintain a rich behavior space. This issue was discussed further in Chapter 3 in connection with the dimension of versatility.

The second law of thermodynamics postulates that most systems tend to decay over time. Yet an intelligent system is one that counteracts this trend. According to the Canadian philosopher Patricia Smith Churchland, an intelligent system is one that utilizes the information it already possesses, and feeds on the energy flux passing through it to increase its store of information.[11]

The entropy law stipulates that all isolated systems degenerate. Hence an intelligent system must be designed to maintain an open boundary to acquire nourishment through order or structure. The structure may take physical form, such as useful energy, or more abstract form such as technological know-how.

## Determinism

Determinism refers to the predictability of a sequence of events. If a deterministic relationship exists between two actions or conditions, then one can be predicted from observing the other. However, if the relationship is only probabilistic, then the one action may lead to the second event or to yet another outcome. Since determinism relates to the relationship among events occurring over time, it is a special aspect of disorder—or its mirror image, structure—in the temporal context.

Often, determinism is a relative concept in that predictability depends on the system or context. When a child throws a model airplane into the breeze, he is unable to predict its trajectory. However, a sophisticated computer model that takes into consideration various factors such as the plane's shape, initial velocity, wind movements and other elements should be able to predict the expected trajectory with fair accuracy. In general, there is little to be gained in attempting to ascertain whether a particular process is completely determinate or inherently probabilistic.[12]

In the nervous system, activation of one neuron by another is probabilistic. For most neurons, a single excitatory impulse will not suffice to raise the membrane potential of the post-synaptic fiber from its quiescent voltage to threshold potential. For example, a single impulse in a motor neuron releases a voltage of about 0.5 millivolt; but the threshold potential for excitation is between 15 and 25 millivolts.[13]

These values imply that at least 30 signals must impinge on a neuron at the same time in order to initiate an action potential. In general, a neuron may possess thousands of synapses, and several hundred of them may "fire" simultaneously (or closely enough to result in the summation of the input). A further complication results from the fact that the effect of an excitatory synapse can be neutralized by the activation of an inhibitory synapse. In summary, the outcome of an excitatory input depends on other independent events and is therefore a probabilistic phenomenon.

Protein synthesis is regulated through various mechanisms, among them induction and repression. The induction of an enzyme occurs when the chemical products of the enzymatic activity stimulate the synthesis of the catalyst by enhancing the transcription of genes in deoxyribonucleic acid (DNA) that code for the enzyme. Repression is the reverse process of inhibiting enzyme synthesis. The information contained in genes is deterministic in the sense that a given gene always yields the same information. But the extent to which the information is transcribed is probabilistic. We cannot predict precisely the amount of protein which will be synthesized, even when we understand the mechanisms of induction and repression.[14]

Several factors come into play in the probabilistic process of synthesizing protein. Only a fraction of the information contained in the DNA of a given cell is utilized and transcribed for protein synthesis in the course of routine activity. The first stage of *transcription* refers to the production of messenger ribonucleic acid (mRNA), while the second stage of *translation* denotes the interpretation of the mRNA by transfer ribonucleic acid (tRNA) to form protein. The identity of the genes transcribed is probabilistic, depending on the ability of RNA polymerase—an enzyme that permits DNA transcription—to bind to a specific region of DNA, thereby inducing protein synthesis. Several mechanisms regulate the synthesis of mRNA by preventing or facilitating this binding.

For example, consider the production of histidine, an amino acid required for protein synthesis. Bacteria grown in a histidine-free environment can synthesize histidine by employing ten different enzymes. But if histidine is added to the bacterial medium, the cells stop synthesizing the amino acid and instead utilize whatever is available in the environment. The fabrication of histidine is prevented when the amino acid binds to a repressor protein; the latter in turn blocks the transcription of the genes coding for enzymes to catalyze the production of histidine.

An example of enzyme induction is found in the synthesis of galactosidase, an enzyme that catalyzes the splitting of lactose into glucose and galactose.[15] In the absence of lactose, bacterial cells synthesize very little galactosidase. If, however, lactose is added to the medium, it induces the rapid production of galactosidase. Lactose, or any inducer molecule, binds to the repressor molecule, effectively preventing the latter from binding to the gene and inhibiting protein synthesis. The gene that codes for synthesis of galactosidase can begin to make mRNA and produce the enzyme. In this way, induction results from the inhibition of an inhibitor.

The firing of a post-synaptic neuron is a deterministic event. When a nerve impulse reaches the end bulb of an axon and is conducted across the synapse, the firing is predetermined. If we know the strength of the electrical impulse at the synapse, we can predict whether or not it will fire, according to whether it is above or below the threshold necessary for excitation of the second nerve.

Under conventional conditions, predictions concerning the traits a child will inherit from its parents are nebulous at best. For example, we can only guess whether the baby will be male or female. Similarly, although mothers over thirty-five years of age face an increased risk of bearing children with Down's syndrome, we cannot predict with certainty that a thirty-nine-year-old mother will have a child with the disease. Hence it is likely but not definite that an older mother will have a child afflicted with Down's syndrome.

In biological evolution, randomization is a critical factor in transmitting traits from one generation to the next. Specimens with differing characteristics are produced, not only as a deterministic outcome of parental traits, but through a partially randomized procedure. The randomization in individual genes as well as in their aggregation into chromosomes may occur through factors such as thermal noise or cosmic radiation. This randomizing factor produces a diverse stock of specimens, which may then be tested for fitness against the environment.

In a similar way, randomization may be used as an effective strategy in computational applications. For example, a systematic procedure known as depth-first search may cause a reasoning system to embark on a search path of infinite length.

This problem sometimes arises in recursive programs: an inquiry that is incompatible with the existing knowledge base can cause a procedure to call itself at every stage in a futile attempt to generate a solution. By introducing a randomizing element in the selection of the subsequent step at each decision point, the system may be able to obtain the desired result in finite time.

The question sometimes arises whether a system is actually probabilistic, or only apparently so due to our lack of comprehension. For example, when tossing a coin, we usually consider the outcome of a "head" or "tail" as a probabilistic event; but a sufficiently detailed model that incorporates the initial velocity, angle, environmental conditions, etc., may well be able to predict the outcome.

Determinism is a perennial—and often unwelcome—consideration in designing artifacts. The specifications for a system may be precise and deterministic; but that may not be the case for the actual outcome. For example, the design requirement for a motor may specify a power output of 100 watts, but the actual value may be lesser or greater due to unpredictable factors in the production process or the materials of composition.

Some of the most intractable problems for engineered devices occur in the post-production stage. Issues such as maintainability and serviceability are partly deterministic: for example, it is possible to specify a fixed maintenance schedule and to design subsystems for rapid replacement. However, other issues relating to availability, reliability, and product life span are more difficult to control or even to predict. In fact, often the most useful model for their "prediction" is based on the simplest assumption: the probability of failure within a given time interval is the same as for any other interval of the same duration. This premise gives rise to the exponential model of failure, which is helpful for modeling characteristics such as the longevity of light bulbs.

The uncertainty associated with a system's activities increases the need for information processing. This uncertainty may arise from the environment or within the system itself.

In the realm of organizational design, a system may deal with the need for information processing by increasing its computational capabilities, or by reducing the need for such activities.[16] The first strategy may be implemented by developing vertical information systems such as policies, standard operating procedures and corporate computer systems; by creating lateral relations such as task forces, matrix organizations, and *ad hoc* meetings; or by acquiring additional resources.

The second strategy of information reduction may be implemented through external activities such as structuring a negotiated environment,[17] or internal tactics such as developing self-contained task modules.

## Realtime versus Batch

This pair of classifications relates to the time involved for responding to external stimuli. Realtime indicates that the response occurs immediately, with no appreciable time lag. An example is found in depositing a check into a bank account and having the funds credited immediately to the account.

In contrast, batch processing refers to the situation where information on a number of events is gathered for analysis at some future point. Returning to the check depositing example, suppose that the check is not processed immediately, but rather put aside with other checks. If this bundle of checks is deposited later, then the process is performed in batch rather than in realtime.

## Timeliness

Information loses value over time due to changes in internal or external factors. An example of an internal disturbance is drift in a system parameter, such as positioning error in a machine tool due to mechanical wear. Examples of external disturbances are changes in temperature or vibration levels. Sometimes the external and internal factors, while innocuous in isolation, may have drastic consequences when they interact. This situation is exemplified by the phenomenon of resonance where external disturbances occur in step with the frequency of internal vibrations. In this case even small disturbances can ultimately generate wild gyrations leading to the catastrophic failure of vehicles, bridges, and other objects.

In a dynamic situation, a purposive system must react in time to adapt to the changes. An intelligent lathe, for example, must respond quickly to tool breakage if it is to avoid ruining the workpiece. Hence the information must be available quickly to allow enough time for the system to respond. The response lag may be due to both software and hardware effects:

- The decision-making unit needs time in which to process the incoming data and make a decision.
- Due to inertial effects, the system needs time to accommodate itself to the new decision.

Time delays can be categorized into different functional phases in terms of sensing, reasoning, and acting stages.[18] Figure 6.2 indicates that *sensing* delays occur in extracting information from the environment. The sensor input is utilized by the *reasoning* subsystem, which consists of the analysis and synthesis phases. The *analysis* of data may be decomposed into *reduction*, *interpretation*, and *comparison* stages. Often, the rate at which incoming information is sampled could be too high for the data to be fully utilized. Moreover, the data is highly redundant both spatially (e.g., continuous regions of a single shade of gray for visual input) and temporally (i.e., the environment does not change abruptly every fraction of a second). Hence the data must first be reduced through techniques such as runs encoding, moving averages, or moving medians. A scheme such as median value extraction is resistant to noise, and will therefore be highly appropriate for certain applications; other techniques will offer differing benefits.

In general, the reduced data must then be combined and interpreted as some higher-level object recognizable by the system. For example, differences in shading may indicate either an edge of a box or a color variation on a single side. The interpreted information must then be compared against the functional requirements to determine how well the system is performing.

After sensing, the next and final major stage is *synthesis*, or the formulation of a

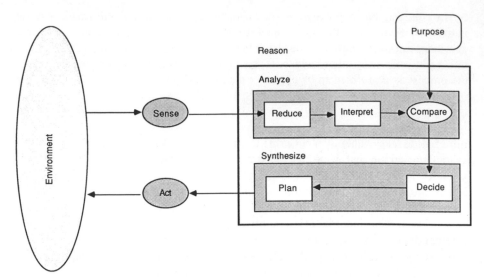

**Figure 6.2** Temporal requirements for sensing, processing, and execution.

strategy for action. This stage may be partitioned into the *decision making* and *planning* phases. The *decision* phase represents the selection of a short-term goal (e.g., increase the temperature by 5 degrees and lower the humidity by 10 percent). A *plan* of action must then be generated to attain the short-term goal. Finally the plan is implemented in the *acting* stage, which results in an observable change in the system or the environment.

A simple example of a system that is susceptible to temporal delays is an automated inspection system. The robotic system must first sense the samples arriving on the materials transport system. This information is processed and compared against a canonical state stored in memory. If the match is good, the sample is routed to the "Accept" batch; otherwise it is sent to the "Reject" pile.

The interarrival time between samples constrains the maximum time available for each stage of the inspection function. Moreover, the time available as a sample moves from the sensing area to the "Accept"/"Reject" decision point defines a window of opportunity for the inspection system. If the robotic system is too slow, it fails to properly discharge its inspection role.

In the biological sphere, the initiation of muscle contraction by the central nervous system occurs roughly in realtime. The signals emitted by the brain travel, via electrical impulses, along nerves directly to the muscle which is supposed to contract. Once the signals arrive at the muscle, contraction occurs almost immediately. There is little delay at any of the three stages of signal initiation, conduction, and motor response.

There is, however, a calculated delay in the body's responses to hunger and thirst. The utilization of energy occurs continuously in the body; fuel is burned continually to maintain homeostasis. For example, consider the process of temperature regulation. Heating and cooling of the body require energy input. By extension, the body's stores of water and nutrient are also constantly being depleted, since these raw materials are being converted to energy to fuel body processes. Yet we are not continuously aware of hunger and thirst. The centers in the central nervous system controlling these cravings must receive signals of the depletion on a continuing basis, but hold the information at bay. This information is then processed in batch, and the response initiated, i.e., eating and drinking. The adaptive trade-off is clear: we forego constant replenishment of the energy stores in order to spend time on other activities such as hunting, sleeping, or contemplation.

The choice of realtime vs. batch processing depends on the application domain; sometimes there is no choice. For example, a control system for an airplane must process information and make decisions in fractions of a second in order to maintain flight. This is especially true of a plane which is inherently unstable in supersonic flight. On the other hand, the data acquisition system for a reconnaissance aircraft may be able to store away the images for batch processing at a later date. In a similar way, a machine tool in an automated factory must monitor its workpiece in realtime; but its job completion data may be stored for future transmission to the factory's management information system.

## Short-term versus Long-term

The debate over short-term versus long-term objectives encompasses two sub-issues relating to reaction times. An organism can react to conditions either quickly or slowly, depending on the urgency of the situation. The characteristic times involved depend on environmental circumstances. If the environment changes rapidly, for example, then the adaptive system must also respond promptly to meet its objectives. On the other hand, knowledge of impending change alone does not require short term response, as long as the change is not imminent.

In the human body, a foreign substance usually undergoes metabolic transformation because the metabolites that are created in the process are less damaging to the system in the short-term than the original substance. However, the long-term effects of the metabolites may be more detrimental to the system. Ethyl alcohol provides one example of this trade-off between short- and long-term consequences.

Alcohol is metabolized in the liver to hydrogen and acetaldehyde. This breakdown occurs in an effort to avoid intoxication, which brings with it a variety of central nervous system malfunctions, such as sluggishness, erratic behavior, and even loss of consciousness. Over time, however, accumulation of the two metabolites can cause considerable damage to the liver itself.[19] Acetaldehyde inhibits the conversion of glucose into usable energy in mitochondria. Hydrogen causes the accumulation of fat, which in turn damages the liver cells. Thus, although alcohol metabolism provides some short-term benefits, it can cause serious long-term damage to the system.

The three types of human muscle fibers illustrate the short-term/long-term trade-off between power and efficiency. Some types of fibers are recruited for tasks requiring short, powerful bursts of tension, while others are more suited to tasks demanding less power but more endurance. The three types of muscle, in terms of increasing efficiency, are glycolytic fast, oxidative fast, and oxidative slow.[20]

*Glycolytic fast* fibers generate the most forceful contractions, but can only do so for short periods of time; they combine a high capacity for glycolysis with the ability to utilize the energy from glycolysis to contract rapidly. Only two percent of the energy contained in glucose is converted to usable energy in glycolysis. Efficiency is sacrificed for power, as the glycolytic fast fibers use up large quantities of glucose and fatigue rapidly. These muscles are employed when we lift weights or leap over puddles.

A second type of muscle fiber, the *oxidative fast* fiber, plays a major role in more sustained periods of activity, such as jogging or swimming. These muscles incorporate mechanisms for aerobic respiration and for rapid energy consumption; thus they can convert glucose to energy fairly efficiently and utilize it to generate muscle contraction rather rapidly. Aerobic respiration converts 39 percent of the stored energy in glucose into biochemically useful forms; hence it is almost twenty times more efficient than glycolysis. Oxidative fast fibers lie in the range between glycolytic fast and oxidative slow fibers with respect to power generated and resistance to fatigue.

*Oxidative slow* fibers are the slowest of all. Their contractions generate the least power of the three but are most resistant to fatigue. They tend to be recruited when activity is to be sustained for long periods of time—such as muscles in the back and legs used for maintaining posture.

The relative importance of power versus efficiency varies with time when muscles are called upon to sustain a given level of performance. In the short-term, power assumes primary importance. The activity takes too little time for the muscles to tire, and fatigue becomes less important. As the physical activity is sustained, however, efficiency in fuel consumption assumes greater importance.

In a contrasting application, a mobile robot must be able to respond to certain stimuli in realtime. However, its finite computing power entails certain delays. The operant question is not "*Is* the response realtime?" but rather "*How* realtime?"

The sensing mechanism may consist of vision for detection under standard lighting conditions, infrared for sensing body heat, sonar for short-range perception, and smell for detection. The high-level control mechanism for the robot is coincident with the reasoning system. Since the reasoning stage has been amply described a few pages ago, we will forego its reiteration here. The low-level control functions are tightly coupled with hardware—in particular, the receptors at the sensor stage and the effectors at the acting stage.

The acting component consists of four subsystems: mobility, manipulation, power, and communications. The mobility function may be supported by wheels, tracks, legs, or some combination thereof. The manipulator may consist of some suitable appendages, or may be nonexistent for a sentry robot. Power may be provided by electromechanical, hydraulic, or pneumatic systems. The robot may communicate with a central station through radio or television; voice capability may be indicated if the robot is to synthesize speech, or if a guard at the central monitoring station is to communicate remotely with potential intruders.

An endangered animal keeps running to flee from its predator, despite the burden of an injured limb. The strain on the injured limb may lead to permanent long-term damage, but flight is necessary to insure short-term survival.

Short-term and long-term are two opposing sides of a trade-off. In this sense, short-term means investing in the near rather than the far future. Depending upon the circumstances of the investment, these can be mutually exclusive. Should an information-processing system be designed to operate at 80 percent of full capacity for a longer period of time or 100 percent for a shorter period? Which design will maximize the cumulative output for the time frame?

Performance and longevity, or short-term versus long-term considerations, tend to be opposing factors in system design when other criteria such as cost or effort are fixed. For example, an engine can be designed to run at high speeds but the resulting increase in stress would hasten the onset of strain, crack propagation, and other failure modes.

In the fall, squirrels gather and eat nuts. However, instead of consuming all the fruits of their labor immediately, they stockpile them for the winter months, when food is relatively scarce. In this way, the squirrels sacrifice short-term benefits for long-term welfare.

This same issue finds many examples closer to our own experience. Short-term versus long-term trade-offs occur, for example, in many economic applications. At the microeconomic level, a teenager of modest means can decide whether to enter the labor force full-time, or to forego current consumption to attend college and thereby enhance the expected value of future earnings.

In a similar way, the competitive disadvantage of the United States in world markets in the 1980's has often been attributed to myopic policies. This is true of consumers, who demand gratification today rather than saving for the morrow. It is also true of producers, who insist on quarterly profits at the expense of long-term investments such as capital expenditures and basic research.

## Summary

Intelligent systems operate over time, involving processes and exhibiting actions that seemingly defy the universal tendency toward disorder. In pursuing their goals, intelligent agents draw on the structure or energy available within themselves or the environment. To sustain effective levels of performance, the agents must incorporate temporal trade-offs, sacrificing short-term gain for long-term benefits where necessary.

- The interplay of static and dynamic conditions is a common theme in system design. A static property can help or hinder a dynamic one, and vice versa.
- An isolated system, closed to the exchange of matter and energy with its environment, tends to decay over time.
- Most physical systems, being open to matter and energy flows, have the potential to enhance their order or structure. The increase in order may be precipitated by natural causes or induced by external agents.

- Disorder refers to the lack of predictability between two items separated by space or time. Determinism relates to predictability between two things separated in time.
- A system must respond in realtime if a delay would entail penalties. Otherwise it may process information in batch mode for subsequent action.
- Expendable resources may be utilized in the near term or invested for use over the long term. The optimal allocation of resources depends on the relationship between cost and benefit over the planning horizon.

# 7

# Process

*The survival of an organization depends upon the maintenance of an equilibrium of complex character in a continually fluctuating environment of physical, biological, and social materials, elements, and forces, which calls for readjustment of processes internal to the organization.*[1]

Chester Irving Barnard

*In a completed science, the words "mind" and "matter" would both disappear, and would be replaced by causal laws concerning "events."*[2]

Bertrand Arthur William Russell

A *process* is a collection of activities or events. An event, in turn, may be defined as some occurrence in space over some duration of time. An event may be viewed as an atomic phenomenon, such as striking a match or recognizing a friend. Or it may be viewed as a compound object consisting of smaller events such as moving a match, grazing it against the side of a matchbox, and drawing it away. In the words of the English philosopher Bertrand Russell:

> Everything in the world is composed of "events". . . something occupying a small finite amount of space-time. If it has parts, these parts, I say, are again events, never something occupying a mere point or instant, whether in space, in time, or in space-time. The fact that an event occupies a finite amount of space-time does not prove that it has parts. Events are not impenetrable, as matter is supposed to be; on the contrary, every event in space-time is overlapped by other events. There is no reason to suppose that any of the events with which we are familiar are infinitely complex; on the contrary, everything known about the world is compatible with the view that every complex event has a finite number of parts.[3]

This book adopts the view that intelligence is tantamount to behavior. Since any behavior is a sequence of events, intelligence depends entirely on the notion of process. The type of processes that concerns us here, of course, takes expression through some physical objects.

In the design of an engineered system, a process is related to the set of activities needed to fulfill some objective. Hence an appropriate process for a system is defined by functional requirements of the system, as is its structure.

## Initiative

Initiative refers to the volition with which a system activates various processes. The contrasting factors are passive and active. A process is *passive* if it is initiated directly by some external stimulus. On the other hand, an *active* process is one that results from the deliberate action of some agent.

Systems are not always passive recipients of environmental stimuli. In fact, a common characteristic of an intelligent system is its ability to manipulate the environment to achieve its goals. Hence, purposive systems may be classified by the degree of initiative they exhibit:

- *Passive*. The system does not operate on the environment in response to input.
- *Active*. The system operates on the environment in response to input.

These opposing factors are useful for purposes of discussion but describe limiting cases on a continuum of passivity; most physical systems will lie somewhere in the middle. Moreover, certain aspects of a system may be active while others are passive. This is true of a sentinel robot whose mission is to observe a phenomenon without interfering, except to the extent required to safeguard its own welfare.

In addition, the degree of initiative will depend on the way in which the system boundaries are defined. Consider, for example, a software package that logs transactions by making changes in a data set or data file. If the package is considered to consist of the algorithmic procedure as well as the data set, then all changes are internal to the system; hence the package is passive. On the other hand, if the data set is part of an autonomous database, then the data set is external to the logging package; consequently the transaction package is an active system.

Some substances contained in the blood stream can penetrate brain tissue actively, while others are transported passively. Unlike capillaries in other body regions, those in the brain prevent many types of molecules from reaching brain tissue by simple diffusion. Brain capillaries do not have any gap junctions in their vessel walls. Thus, only very small molecules can diffuse freely in and out of the vasculature.

Some amino acids and ions rely on carrier-mediated active transport systems for movement from blood vessel to extracellular space. These transport systems tend to be very specific. For example, potassium ions appear to be actively transported from the brain's extracellular space to the blood. Hence, potassium concentration in the extracellular brain fluid tends to be much lower than that found in blood plasma.

On the other hand, certain substances are transported *passively*. Lipids can pass through the capillary endothelia with relative ease. For this reason, the transport of lipid-soluble substances such as carbon dioxide, oxygen, and certain drugs, occurs fairly rapidly.

Active and passive mechanisms offer differing advantages and limitations. The use of active transport mechanisms instead of passive diffusion provides extra protection for the brain from potentially harmful fluctuations in blood composition. But active transport takes extra time and energy. Passive diffusion is the method of choice when reliable transport is desired.[4]

The movement of ions and food illustrates the integration of active and passive transport. The voltage difference across a neural membrane is maintained in part by a "sodium pump." In its quiescent state, the cell membrane of a neuron acts to maintain a concentration gradient of sodium ($Na^+$) ions, using energy to actively transport the $Na^+$ ions out of the cell. The membrane is selectively permeable to the $Na^+$ ions; in its resting state it does not permit their reentry when the membrane depolarizes. However, during an action potential, it suddenly becomes permeable and the $Na^+$ ions diffuse inward passively but rapidly, since their concentration outside the cell is significantly higher.

The transport of food down the esophagus and through the intestine provides another example of an active process. Once food has passed the throat, waves of muscular contractions—peristaltic waves—transport the food mass through the digestive tract. These regular contractions also occur in the small intestine, propelling the digested food and enabling absorption of nutrients through the intestinal walls. Thus, both the esophagus and small intestine act upon the food mass to direct its movement.

Another facet of active and passive transport is found in the movement of ions in digestion. When hydrochloric acid is secreted during digestion, either active or passive transport takes place, depending on the direction of the chloride ions. The secretion of the acid, which dissolves food particles and kills microbes in food, is effected by the parietal cells in the epithelia of the stomach. The secretion of hydrogen ions involves active transport while chlorine ions depend on facilitated diffusion as well.

In the parietal cells, water molecules are split up into hydrogen ($H^-$) and hydroxyl ($OH^-$) ions. The hydrogen ions are then actively pumped out of the cell into the stomach lumen. The energy for this active transport is provided by the breakdown of adenosine triphosphate (ATP) into adenosine diphosphate (ADP). The separation of a phosphate group from ATP provides the energy for forcing the hydrogen ions across the plasma membrane.

Meanwhile, chlorine ions are first obtained actively from the bloodstream, then released passively into the stomach cavity. The active absorption of these ions into the parietal cells occurs in conjunction with the excretion of bicarbonate ions to maintain charge distribution. The energy for this transport also comes from the breakdown of ATP. In the second phase, the chlorine ions leave the cells passively by way of carrier molecules which require no energy consumption. Hence, the secretion of hydrochloric acid in the stomach depends on a combination of both passive and active modes of transport.[5]

A final illustration of passive versus active systems can be found in the simple comparison of two timepieces. A sundial is a passive monitor of time. It is a highly reliable device, having no moving parts to break down; unfortunately, it is useless under clouds, indoors, or at night. A clock is an active timekeeper that can operate without sunlight; but it is wont to break down, wear out, or deplete its energy reserves.

## Direction

A process may be viewed as a sequence of events that transpire in one direction or another. The dichotomy between forward versus backward relates to the selection of the starting condition or the final state as the initial focus of interest.

The interplay between enhancement and reduction refers to transformational operations on data, whether in terms of accentuating separable features or distilling them into a smaller set of parameters.

## Forward versus Backward

Information processing in the *forward* direction denotes the situation where a stock of initial information is evaluated to determine its consequences. If a system proceeds in a forward direction, it performs tasks and processes information with a general awareness of the criteria for determining success, but without constant guidance from any particular "destination" or objective. On the other hand, if an agent is designed to proceed *backwards*, the system begins with some goal or final destination in mind and then attempts to discover a route that leads logically to the facts of the case.

An analogy lies in searching for a pizza parlor. I can leave home and drive around the neighborhood until I find a pizzeria, or have one in mind and determine how best to get there. The former strategy is forward processing, and the latter strategy backward.

Several other pairs of terms related to this dichotomy are sometimes employed by writers in management, psychology, and computer science. In design, *top-down versus bottom-up* refers to the direction of progression in the making of an intelligent system. The architect can begin with a high-level goal in mind, and then devise the necessary machinery or design to fulfill his objectives; this is known as a top-down strategy. On the other hand, the designer can also come up with various parts or building blocks, without envisioning some goal. Then, when many "parts" have been designed, he can assemble them into a higher level system. So proceeds bottom-up design.

Related to forward versus backward is the concept of *fact-driven versus goal-driven reasoning*. If the system's only guidance comes from the information that is input, then it will exhibit fact-driven reasoning, and thus will operate in the forward direction. If, however, the system begins operation with some goal in mind, then it will proceed "backwards" because its reasoning and information-processing will be guided by this goal.

When discussing learning and conditioning, the terms *push* and *pull* can also be used to describe the conflicting sides of this dimension. In learning, a teacher can either guide a student by directing or "pulling" him toward the desired concepts or behavior through rewards, or by "pushing" him forward with negative reinforcements.

In conditioning, the lure for enticing the subject to perform the desired action is often the reward which he receives after performing "correctly." Conversely, the punishment acts as the instrument for "pushing" away any undesirable behavior. This process of reward and punishment is used by animal trainers and somewhat more subtly by teachers of human pupils. In an analogous way, the production at each station in a factory can be pulled forward by a request from the subsequent station, or pushed forward by workpieces accumulating in its input queue from the preceding station.

In the physiological arena, vision illustrates the way in which data is processed in a forward-chaining fashion, from initial stimuli to final result. Light impinging on the retina is detected by cells called rods and cones. These cells synapse with second-stage cells, such as bipolar cells, which in turn stimulate ganglion cells. The layered structure serves to aggregate redundant information and accentuate interesting features in the visual field.

The axons of the ganglion cells form the optic nerve, which connects the retina with the brain. The optic nerves from both eyes converge at the optic chiasm, located in the base of the brain. There, some of the nerves cross to the opposite side of the brain, ensuring that each hemisphere obtains at least partial input from both eyes.

The optic fibers lead directly to many regions in the brain, but most enter a relay center in the thalamus known as the lateral geniculate nucleus. This relay station disperses the optic fibers into several directions. Some nerves are conveyed, for example, to the brainstem and cerebellum, where the information is used to coordinate the movement of the head and eyes. Most of the fibers from the lateral geniculate nucleus are passed on to the visual cortex. But the cortex itself has differentiated regions. Different areas of the visual cortex are responsible for processing binocular information, interpreting color, detecting movement, and other functions. The processed signals then travel by different pathways, depending on the type of information (such as object recognition or spatial location) to visual association areas in the frontal lobe.[6]

The visual apparatus embodies a forward-processing approach since each stage of transmission is concerned solely with the task of accepting information and passing it on to the next stage, after enhancing it in some way. At the neuronal level there is no conception of the overall task, only the mission of conveying information in the form of electrical impulses. This information is gradually coalesced through a series of integrative mechanisms, ultimately forming the basis for feature recognition and decision making.

A laboratory rat demonstrates a backward-chaining process when it learns new behavior through operant conditioning. The rat is placed in an experimental environment consisting of a box with enough room for it to move about comfortably. From one wall protrudes a lever. If pressed, the lever opens a door that releases a food pellet as reward. Although at first the rat will only press the lever occasionally and at random, it quickly learns that it will receive the reward by doing so. Thus it "learns" a new behavior pattern. In so doing, the rat is exhibiting a backward chaining process. A perennial goal of a hungry rat is to obtain food. In order to obtain food, the rat must press the lever and thereby learn a procedure for achieving its goal.

An example of forward processing is the use of data-driven reasoning, in which information is used to derive consequences. For example, the syllogism "Robots have sensors" coupled with "Robin is a robot" results in the conclusion "Robin has sensors."

In contrast, backward processing is exemplified by goal-driven reasoning. In this approach, a hypothesis is first formed, then an attempt is made to determine whether the facts support such a conclusion. In the preceding example, the hypothesis "Robin has sensors" would be tested by determining whether the knowledge base supports such a conclusion; in particular, the knowledge "Robots have sensors" and "Robin is a Robot" would validate such a hypothesis.

*Advantages of Forward versus Backward.* Forward processing is preferable to backward when the former yields results more readily than the latter and vice versa. Forward movement is appropriate, for example, when the number of alternatives at each decision point is fewer when moving in a forward rather than a backward direction.

Let us return to the pizza parlor analogy. If my neighborhood is chock full of pizzerias of equal caliber, then I can simply jump into my car and drive forward without too much concern for the final destination. Moreover, if I am new to the neighborhood and have no clue where these critical oases lie, then I also have no choice but to get in the car and start driving. On the other hand, suppose that pizza shops are few and far between in my town. Then a better strategy would be to first find out where a pizzeria is located, then seek a backward route from the store to my place.

To validate a theorem in mathematics, it is often more efficient to begin with the theorem as a hypothesis and then determine whether it is supported by the known facts. This is the basis for the refutation procedure used in the logical programming language of Prolog.[7]

Consider the maze that presents a special fascination to school children. The number of forks, or branching factors, at each decision point is usually larger in maze problems when moving forward than when moving backward. This is not by accident, since the games' creators expect players to begin from the "Start" position rather than the "End" location. Hence a shortcut to the problem is to begin from the "End" position and move backward.

Often the best strategy is to use some combination of forward and backward processing. The diagnostic procedure used by physicians, for example, relies on initial observations to generate a handful of alternatives in a rapid process of data-driven reasoning, followed by attempts to validate the hypotheses through backward reasoning. If all the hypotheses fail to hold, then a new set of hypotheses is generated and the backward procedure begins anew.

Of course, professionals are not the only people who employ this technique. Students quickly discover that the most rapid way to respond to the problem "Derive the formula $G$" in a physics quiz is to employ a bidirectional procedure, much like laying railroad tracks from both New York and San Francisco until they meet somewhere in the Midwest. The student begins from the goal $G$ and attempts to "unravel" the variables into a recognizable set of factors. When she reaches an impasse, the student moves forward from the premises and attempts to massage the variables into the intermediate state of the equations resulting from the backward procedure. The forward and backward modes of reasoning alternate until the two lines of thought are made to join in a seamless train of argument, from the initial premises to the final goal.

## Enhancement versus Reduction

Reduction versus enhancement of information addresses the types of processing that may be applied to input data. Reduction refers to the compaction of information in order to highlight the salient features; in other words, defining the forest despite the trees. For example, the incomes of the residents in a town might be reduced into summary figures relating to the mean and standard deviation of incomes.

Enhancement is the converse term for expanding upon key features despite a dearth of information. This phenomenon refers to the reinterpretation of information to highlight critical aspects and suppress peripheral properties. Enhancement can only refer to the apparent increase in knowledge or information. Information corresponds to order or structure. Thus, if information were actually created, the second law of thermodynamics would be overturned—an unlikely event.

Any sensory organ filters information by default, taking in at a given time only part of what is available. In the case of the eye, the image from the environment is rich in information due to its largely continuous nature along two dimensions: the spatial distribution of points and the spectral pattern of light emanating from each point. As the signal reaches the retina, the spatial distribution is condensed into a set of discrete sensors in the form of rods and cones; and the wealth of spectra into the three primary types of color detectors. In this way, the output signals from the retina reflect only a minute fraction of the input image.

The determination of boundaries is one of the basic problems in visual pattern analysis. In order to highlight boundaries, impulse transmission rates increase in certain regions of the photoreceptor cells, and decrease in others. To produce this effect, the photoreceptors utilize the mechanism of lateral inhibition. They excite the ganglion cells directly beneath the beam of light (and therefore beneath the responding photoreceptors) and simultaneously inhibit ganglion cells beneath adjoining photoreceptors.

This scheme is illustrated in Figure 7.1, which depicts an object consisting of a light and a dark region. The incident rays from the lighter region have an intensity level of 10, which is twice the level of 5 from the dark region. The photoreceptors relay these signals directly ahead to the ganglion cells by way of transmitters, and obliquely forward by way of inhibitors. In the left half of the figure, a positive transmission rate of 10 pulses per second is associated with a lateral inhibitory rate of 2 pulses; these values are halved in the right side.

The ganglion cells sum the positive contributions from the transmitters and the negative inputs from the inhibitors. The resultant signal has a level of 2, 3, 6, or 7 pulses, as shown in the bottom row of the diagram. This process accentuates the boundary between the light and dark (output levels of 2 and 7) over that of the regions themselves (levels of 3 and 6). Thus, lateral inhibition enhances the contrast in firing rates at the edges of the stimulus, which in turn facilitates boundary detection and pattern analysis. In this way only the most salient features of the environment are accentuated by the visual apparatus.

The same type of enhancement procedure is employed in artificial systems to clarify features in information-poor signals. Such methods are routinely employed, for example, in interpreting acoustic signals in military applications, or in enhancing electromagnetic data from astrophysical sources.

When sensory information needs to be reduced, the gateway to the cerebral cortex in the form of the thalamus serves as a major coordinating center. This section of the brain receives and correlates data from receptors located throughout the body, then sends condensed reports to the cerebral cortex for further analysis

Object

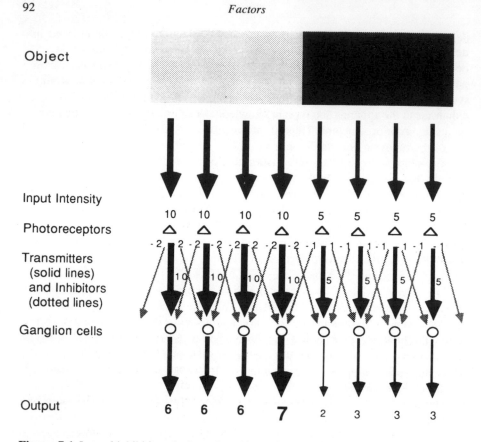

Input Intensity

Photoreceptors

Transmitters
(solid lines)
and Inhibitors
(dotted lines)

Ganglion cells

Output

**Figure 7.1** Lateral inhibition: the intensity of incoming light is passed directly forward as a positive signal, and laterally as a negative signal.

and consolidation of information. In addition to sensory information, however, the thalamus also processes data that bears no direct relation to sensory input. This information comes from the cerebellum and the limbic system, primarily. Thus, the data that the thalamus receives and transforms affects both discrimination of sensation and the integrated activity of structures involved in motor control and emotion.

Neuronal impulses from many parts of the body arrive at their final destination, the thalamic nuclei, via connecting cables such as the optic nerve. The input neurons synapse with thalamic neurons, which then project directly into the neocortex. Blending of afferent information—the integration of sensory data with related afferent information—occurs to a significant extent in the thalamus. This blending is further enhanced by the merging of data and the complex feature analysis that takes place in the cortex.[8] The fusion of parallel trains of data allows for the amplification of salient signals and the neutralization of noise or spurious information.[9]

Patches of active membrane in the dendritic tree of some neurons amplify information that has travelled extended distances. These active sites, capable of depolarization, occur at the point of confluence of two or more subbranches of the dendritic tree, and are separated from the neuron's cell body and axon by regions of dendritic membrane. They increase the probability of an excitatory synaptic event's generation of an action potential by boosting the normal post-synaptic (electrical) response to such an occurrence. In this manner, dendritic "hot spots" serve to enhance the information received from distant dendritic inputs.[10]

## Coordination

The components of a system must interact in coherent fashion to yield purposive behavior. The degree of centralization refers to localization or distribution of control. The level of feedback relates to monitoring activity to determine how well objectives are met, and taking corrective action based on any discrepancies between intention and reality.

### *Feedback*

Feedback relates to the monitoring of results by some initiating agent. In a closed-loop configuration, the system issues a signal to activate a process, then monitors the outcome. If the outcome is unsatisfactory, the input signal is modified and sent again in the hope of bringing about a more satisfactory result. The general scheme for closed-loop control is shown in Figure 7.2. A referent signal enters a summing module which compares the desired state with the actual conditions. The discrepancy is fed into a control unit which determines the appropriate response. The ensuing command enters an amplifier which provides the energy to activate the module under control. The actual state of the controlled system is monitored and fed back to the summing element, thereby closing the control loop.

In open-loop control, in contrast, the controlling unit issues a command and expects the system to perform as desired. Since there is no feedback loop, the regulating module cannot obtain any information on the status of the module to be controlled. The regulating unit therefore issues the same command no matter how well or poorly its goals are met by the subordinate system.

Humans have developed a variety of mechanisms to counter the environment in maintaining a relatively constant body temperature. These mechanisms provide an

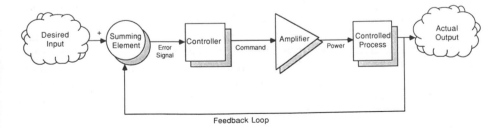

**Figure 7.2** Schematic of a closed-loop control system.

example of a closed loop system. The rate of signal initiation in certain nerves depends on body temperature, so that when the body is losing heat briskly to the environment, these nerves fire rapidly. This sends the message to the brain, which then influences, via the efferent pathway, the rates of muscle contraction. Since muscle contraction produces energy that is dissipated internally as heat, the body can warm itself quickly and efficiently.

Under normal conditions, the reproduction of cells is a closed loop system, correlated with the ratio of surface area to volume. The diffusion rate of nutrients is the limiting factor in the timing of each cell's reproduction. Once the ratio of surface area to volume drops, nutrients cannot diffuse through the cell at a rate fast enough to sustain its survival, let alone growth. Hence, the cell must divide to increase the surface area available for diffusion. Moreover, control of the proliferation of epithelial cells appears to be mediated by intercellular communication. In a growing epithelial sheet, contact among dividing cells appears to stop further cell division; each cell's reproduction is controlled, to some extent, by the environment.

In contrast to normal cell proliferation, uncontrolled growth characterizes most cancer cells. Malignant tumors grow wild, ignoring any rules of differentiation. This type of abnormal growth represents an open-loop process. No feedback for the control of proliferation exists for the tumor, whereas healthy cells do possess such control mechanisms.[11]

The vestibular system located in the skull provides for balance and orientation, including eye reflexes, head movements, and aspects of posture. For example, it ensures that the eyes are fixed on the same point in space despite movements of the head. The vestibular system senses orientation through hair cells which respond to rotational or translational changes in the head. The mechanism is sensitive to both motion or acceleration and displacement or orientation. In addition to providing spatial information for volitional movement, the vestibular system plays a key role in several regulatory reflexes. One of these is the vestibulospinal reflex which opposes undesirable movement. This reflex occurs in the neck muscles to regulate the orientation of the head, and in leg muscles to maintain balance (see Figure 7.3).

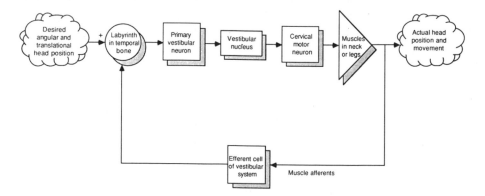

**Figure 7.3** Vestibulospinal reflex as a closed-loop control process. (After Angevine and Cotman (1981), p. 82.)

The reflex takes a closed-loop configuration where the desired position of the head and body, in terms of angular and translational displacement, is transmitted through a cascade of control modules to the appropriate muscles. The actual position is detected by sensor neurons and conveyed to the labyrinth via the efferent cell of the vestibular system. The labyrinth detects any discrepancy between the desired and actual positions, and activates the appropriate control signals to the regulating modules.[12]

In the artificial realm, messages are usually transmitted in open-loop mode. This is true of letters, telegrams, and data packets sent along a communications channel. In the last case, even when the sender obtains an acknowledgement from the intended recipient, it is possible that noise has transformed the original message into a mutant version. Therefore the partial feedback, even if it is received, can be unreliable.

On the other hand, most control systems are of the closed-loop variety. Examples are found in controlling the speed of a car, the altitude of an airplane, or the depth of a cut of a milling machine. Although the initiating command may be sent in open-loop format, the actual value of the state to be controlled is fed back to the controlling unit. For example, the depression of the gas pedal determines the fuel intake and the power output, which eventually translates into vehicle speed. The process so far is open loop. However, the speedometer monitors the automobile's rate of movement and alerts the driver, who ultimately closes the loop to the gas pedal.

For a deterministic system, closed loops are unnecessary: the system may be set on course initially, then trusted to behave as set for all time. However, closed loops are advisable in the face of uncertainty. This uncertainty may be due to stochastic phenomena—such as random disturbances—or incomplete knowledge of the situation—such as the physics of composites fabrication processes.

## Centralization

Centralization refers to the degree of localization for information processing. This dimension may be further resolved into two related facets: (1) the degree of *control*, ranging from autocratic to democratic, and (2) the *impact* of activities on system performance, from localized to global effects. The former may be viewed as a top-down concern, and the latter as a bottom-up effect.

The symbiosis of global purpose and local autonomy characterizes most complex systems, whether in the natural or artificial domain. In the words of a practitioner of the management arts:

> The executive process, even when narrowed to the aspect of effectiveness of organization and the technologies of organization activity, is one of integration of the whole, of finding the effective balance between the local and the broad considerations, between the general and the specific requirements.[13]

Perhaps the simplest control arrangement is found in complete centralization, in which a single center determines all transactions. However, if the system operates in a complex environment, one large center might become too cumbersome. In such instances, the overall system could rely on a network of smaller aggregating centers, each performing its own integrative tasks. On the other hand, working with many small centers might prove to be awkward if voluminous amounts of communication are necessary.

The optimal level of centralization is dictated in part by the required speed of response. When the requisite response time is low, decision making may be decentralized: time is available for democracy, deliberation, and debate among interested parties.

When decisions must be made more quickly, the luxury of debate vanishes. In this case, authority must be vested in a central party: the air traffic controller must oversee the arrivals and departures at an airport; and the general must direct military operations.

At extremely high rates of decision making, the appropriate level of centralization again decreases: a central unit may not be able to receive and transmit information within the available time, or the information may be too voluminous to handle effectively at a single site. Hence the individual soldier must decide whether to dodge an artillery barrage rather than wait for missives from his commander; the airline pilot must decide which way to swerve in the face of an imminent collision rather than await a message from the traffic controller; the reflex mechanism must ensure that an errant foot is jerked away from a thorn rather than await a strategic decision from the brain.

The effective level of centralization as a function of required response speed is depicted in Figure 7.4. The curve initially rises with an increase in the response rate, but eventually declines.

The mere existence of a brain indicates that information processing in humans is, to some extent, centralized. The brain serves as a center for coordinating responses to most of the data received by our sensory mechanisms, and also controls

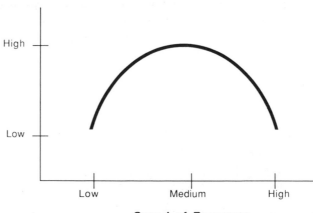

**Figure 7.4** Appropriate level of centralization as a function of requisite response speed.

several "involuntary" functions. However, some decentralization is evident in the organization of the nervous system. The reflex pathway invoked, for example, in the retraction of a hand from a hot stove, never passes through the brain. Instead, the sensory neurons go to the spinal cord, and are connected to motor neurons that originate from the spinal cord as well.

The primary center for control of cardiovascular activity lies in the brainstem medulla. However, auxiliary controllers for regulating heartbeat exist within cardiac walls. As a result, loss of brain function does not necessarily lead to the paralysis of the heart; it can still pump blood.

Nodal tissue, located within the heart, enables that muscle to contract independently of any stimulation by the central nervous system. This tissue, composed of various bundles of fibers derived from cardiac muscle, no longer beats, but has developed the ability to spontaneously depolarize and conduct electrical impulses. The sinoatrial node, located in the right atrium, acts as a pacemaker. Spontaneous depolarization occurs fastest here. The impulses produced by the sinoatrial node are conducted through the walls of the atria causing their contraction. The action potential then makes its way to the atrioventricular node, which passes the impulse quickly through the ventricular muscle to achieve near-simultaneous contraction.

Both our cardiovascular and respiratory systems are largely centralized. The cardiovascular control center is located in the brain stem and regulates blood pressure via information from a network of pressure receptors in both arteries and veins. These baroreceptors are usually found in areas where the walls of the vessels are thinner than usual. The nerve endings of the baroreceptors are highly sensitive to stretch or distortion, and the degree of wall extension serves as a fairly accurate indicator of pressure within the artery. Any increases or decreases in arterial blood pressure are signalled by corresponding increases in rate of firing of the receptors. Once these signals reach the brain, it can respond by varying contraction, strength, and arteriolar or venous dilation.

The control center of the respiratory system resides in the medulla oblongata region of the brain. This region regulates both the volume of air inspired per breath and the rate of breathing in order to maintain the blood levels of carbon dioxide within narrow limits. Although oxygen levels are monitored as well, the primary information concerns carbon dioxide. If the concentration of this gas rises, so do the rate and depth of respiration; conversely, a decrease in the compound leads to attenuated respiration.

Mechanoreceptors in the lungs monitor the degree of inflation in these organs, while proprioceptors in the thorax and abdomen keep track of the contraction levels in the respiratory muscles. The carotid and aortic bodies—located next to the carotid artery and aorta, respectively—both sense levels of oxygen, carbon dioxide, and acidity in the arterial blood. Receptors within the medulla itself are especially sensitive to the acidity of their extracellular environment.

This centralization of respiratory regulation appears to allow for more effective control of input information (e.g., blood gas concentrations) with output (e.g., respiratory rate). The refined control protects the brain and the entire system from wild fluctuations in oxygen supply and carbon dioxide elimination.

Yet another example of centralization in the brain lies in one of the functions of the Wernicke area, which handles the comprehension of written and spoken material. This module interprets signals emanating from the auditory region, visual cortex, and other localized centers in the brain that affect speech production. In this way, the Wernicke area plays an integrative role.

The spectrum of computers available today—from micros to mainframes—represents a fine example of the pros and cons of centralization. Ever since the early days of computing, computer capabilities have continued to expand. This trend has been driven by the following factors: existing computers are never quite fast enough for the largest computational problems; even "large" computers possess miniscule power compared to the human brain; and the retail cost of hardware increases only logarithmically with computing power.

So it came to pass that mainframe computers dominated business applications by the 1960's, while supercomputers handled scientific and technical problems. These devices, being expensive and unreliable, were housed behind glass walls in climate-controlled environments and attended to by humans.

The centralization of computing power had its shortcomings. Although time-sharing technology helped to distribute computational resources to geographically dispersed users, these users were often frustrated by the inefficiencies. Chief among these complaints for commercial applications was the sluggish pace of software development, which often dragged on for years.

Against this backdrop, the minicomputer appeared. This innovation offered only a fraction of the power of mainframes; but it was balanced by much lower cost as well as friendlier interfaces, higher reliability, and lighter maintenance requirements. As with the mainframes, these machines were first employed by technical users but gradually permeated the business realm. By the 1970's, they were a standard feature of the commercial landscape.

For largely the same reasons that mainframes displaced minis, microcomputers came into their own in the 1980's. Personal computers provide computing capability to individual users, while microprocessors provide such capability to individual gadgets (e.g., clock pens) or their subsystems (e.g., automotive engine regulation.)

Today no single technology—whether in micros, minis, or mainframes—dominates the computing market. They are used synergistically to provide the proper mix of centralized or decentralized computation, as required by the specific application.

The spectrum of centralization issues is illustrated by communication systems whose control strategies range from monarchical to near-anarchistic. These range from the star configuration to the ring, bus, and network arrangements.

The *star* topology is one in which each regular agent or node has a single connection to a unique supervisory node (see Figure 7.5). The advantages of this architecture arise from its simplicity: cost-effectiveness due to the minimal number of communication links, and ease of control due to mediation of all communications by the supervisory node.

One disadvantage of the stellar strategy arises from its limited expansibility; the supervisor can accommodate only so many nodes before overloading its computational capabilities. A more acute limitation is found in its vulnerability; when the supervisor breaks down, so does the entire communication system.

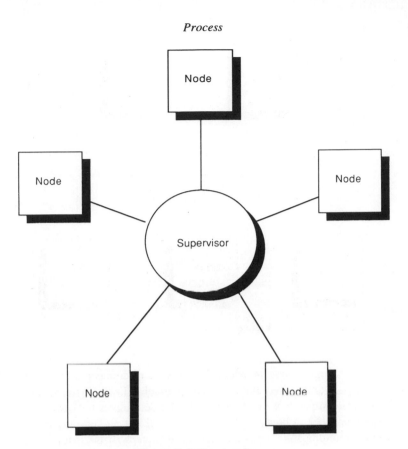

**Figure 7.5** Star topology.

The *bus* topology is one in which all the nodes share a single communication line (see Figure 7.6). In this approach, the control may be centralized—as in the star architecture—or it may not.

The decentralized approach requires more sophistication in internode communications, as two or more nodes might independently decide to begin transmitting into the bus at the same time. An effective solution is the Carrier-Sense Multiple-Access/Collision Detection (CSMA/CD) protocol. A node wishing to transmit monitors the bus until it is quiescent. As it begins transmissions, the node also listens for the possibility of jumbled signals due to simultaneous transmissions from other nodes. If the signals are unadulterated, the node completes its transmission; otherwise it stops, waits for a random length of time, then tries anew.

As with the star topology, the bus configuration is cost-effective. It is also vulnerable to failures in the bus. Further, it has limited capacity for expansion in the number of nodes, due to the finite bandwidth or carrying capacity of the bus. In addition, the CSMA/CD approach limits the physical span of the bus: the longer the bus, the longer a transmitting node needs to wait to ensure that it has a clear channel. Hence, the bus topology is only used for local networks spanning no more than a few kilometers.

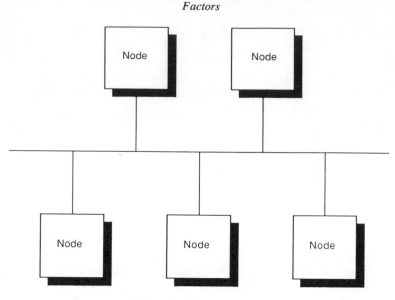

**Figure 7.6** Bus topology.

The *ring* topology is one in which the nodes are connected into a loop (see Figure 7.7). Its advantage over the first two approaches lies in increased reliability. There is no single supervisory node whose health determines the operability of the entire system. Further, a break in a single link will not prevent any of the linked nodes from communicating with another. The decentralized nature of the ring topology requires coordinative strategies more involved than that for the star configuration, but less than for the bus.

In the *network* topology, each node may be connected to one or more neighbors, as illustrated in Figure 7.8. The ring topology is a special case of this type of architecture.

Like the ring, the network approach is usually characterized by decentralized control. The two communication strategies in networks relate to dedicated connections and message modularization. In the first approach, nodes communicate with each other via *virtual line* or *circuit switching*. Here, a temporary connection is established between two communicating nodes. If the two nodes are adjacent, they simply utilize the physical circuit connecting them. Otherwise, they establish a virtual circuit by commandeering the physical links attached to intermediate nodes. This set of links is dedicated as a private circuit throughout the duration of the transmission. The disadvantage of this approach, of course, lies in the fact that other pairs of nodes may be unable to communicate throughout the private "conversation."

A more popular approach that transcends this limitation is *packet switching*, in which a transmitting node breaks up its message into packages of a fixed size. Each package is also given control information such as its position in the sequence of packages as well as the name of the recipient node. Any node that receives such a message passes it on to one of its neighbors. The selection of the subsequent node depends on the location of the recipient node. Since physical circuits are not dedicated to a particular "conversation" between communicating modes, each physical link may be used first by a message packet from one node, then a second, back to the first, and so on.

One advantage of the network configuration is its reliability: the multiplicity of links implies that an alternative path can often be found between two nodes, even when a particular link or node has failed. Another advantage lies in expansibility: adding more nodes usually requires no more effort than constructing more links. For these reasons, the network approach is the topology of choice for large-scale communication systems.

In the business world, companies must be able to adapt their structures in accordance with environmental demands. The experiences of the du Pont Company and General Motors Corporation in the early part of this century illustrate this need for the orchestration of centralizing and decentralizing factors.

At the close of World War I, the du Pont Company intensified the centralized structure which had served so well since the consolidation of its operations in 1902.[14] However, the company had grown dramatically during the war, both in revenues and product diversity. In 1913, fully 97 percent of its business involved explosives. However, a deliberate decision was made in 1917 to diversity into dyes, paints, and other chemical products. Even so, the company retained its traditional structure of partitioning by departments. All products were grouped under a single production department and promoted by a single marketing arm.

Unfortunately, the centralized format that had been effective in earlier times

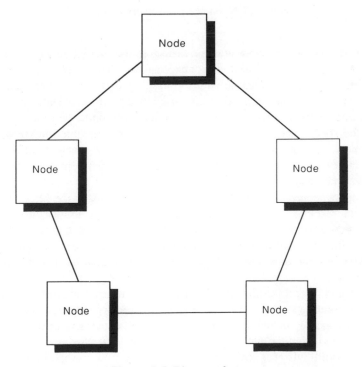

**Figure 7.7** Ring topology.

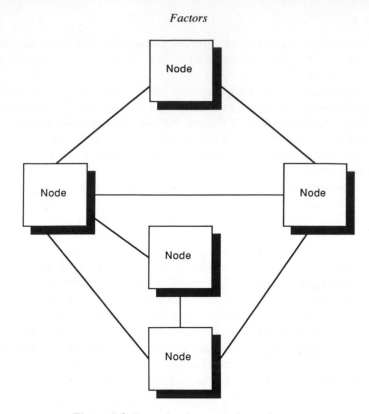

**Figure 7.8** Example of a network topology.

proved inadequate for the new strategic thrust. In the first six months of 1921, the firm lost money on every product except explosives.[15] That September the company was molded into a divisional structure consisting of five relatively autonomous product lines. These operations were supported by a portfolio of staff functions such as engineering and legal services. The divisional partitioning permitted each group to meet the special needs of its environment in a more flexible, timely way.

The new structure served the company well. Losses were transformed into profits, and the basic structure remained intact even when new product lines were added over the ensuing years.

The opposite problem was faced by General Motors at the beginning of the century. Its founder, William C. Durant, had wrought a loose confederation of companies thrust together by acquisition and combination, both horizontally among automotive builders and vertically to their suppliers. Durant was an empire-builder. He cared more for organizational growth and sales volume than for profits or administration.

The dearth of accounting and information structures implied the lack of a coherent marketing strategy and loss of control over expenditures. Sales dropped, cash flow dwindled, and the firm's stock price plummeted. Eventually, Durant was forced to resign from the presidency in November 1920.[16]

A new multidivisional structure for the company was proposed by Alfred P.

Sloan, Jr. This structure was similar to that at du Pont but had been developed independently by Sloan. In his analysis of the corporation, Sloan had envisioned greater centralization and administrative control; but this was to be balanced by divisional autonomy to promote innovation and marketing flexibility.

Each division, such as Buick or Cadillac, was to develop its own identity and pursue its own market segment. The product lines and market sectors, however, were expected to overlap to some extent. These divisions were to be coordinated through financial and information structures, and to share major resources such as research and development.

The new corporate structure was implemented over the four years ending in 1925.[17] It served General Motors so well that, by the 1960's, the firm came to be regarded as a paragon of industrial organization and the epitome of free enterprise.

A key disadvantage of a decentralized scheme is the possibility of incoherent action, even to the point of total inaction. A completely distributed system may be subject to paralysis by the impossibility of resolving conflicts among sovereign agents possessing inflexible patterns of preference. In a straightforward ordering of preferences, it is not always possible to determine the optimal set of choices for a group of individuals. In the words of the welfare economist Kenneth Arrow:

> If we exclude the possibility of interpersonal comparisons of utility, then the only methods of passing from individual tastes to social preferences which will be satisfactory and which will be defined for a wide range of sets of individual orderings are either imposed or dictatorial.[18]

This phenomenon applies whether the system is a society composed of individuals or a device made up of components.

In practice, however, there usually exist ways to compare utilities or preferences among individuals. The entire concept of a pluralist society is based on the selection of some weighted combination of individual preferences. Hence a mechanism for reevaluating preferences and resolving interagent conflicts becomes necessary. One simple way to attain such a mechanism was illustrated for gaining access to a network: the Carrier-Sense Multiple-Access/Collision Detection protocol described earlier.

## Summary

A process is a collection of activities or events. Intelligence, as behavior, takes expression through processes. The design of an intelligent system requires the architect to decide on the levels of initiative, the directionality of activities, and strategies for coordination.

- A system may display any level of initiative, ranging from passive at one extreme to active on the other. Many systems incorporate behaviors that indicate a mixture of initiative.
- A process may be viewed as proceeding in one direction or the other. Some polarized terms in this category are forward vs. backward, also known by other terms such as top-down vs. bottom-up, fact-driven vs. goal-driven, and

push vs. pull. Another parameter in this category is the contrast between enhancement vs. reduction of information.

- The components of a system must coordinate their activities to effect their collective objective. Feedback refers to the monitoring of actual conditions, comparing them against desired results, and taking any corrective action that may be required. Centralization relates to the localization of control required for interaction among the system modules.
- The optimal level of centralization depends, as do many other system attributes, on environmental conditions. Centralization of control is preferable when the entire resources of the system must be marshalled quickly for the system to attain its goals. When the pace of environmental change exceeds the transmission times required for intrasystem communication, a decentralized arrangement is indicated. The most comfortable situation occurs when environmental changes are slow and the system need not orchestrate all its resources simultaneously to fulfill its objectives. In this case, decentralization will allow for efficiency in information-processing operations and may facilitate innovative behavior and systemic learning.

# 8

# Efficiency

*Natural selection will not necessarily produce absolute perfection.*[1]
Charles Robert Darwin

*The natural sciences are causality research; the sciences of human action are teleological. . . . The very category or concept of action comprehends the concept of means and ends, of preferring and putting aside, viz., of valuing, of success and failure, of profit and loss, of costs.*[2]

Ludwig von Mises

*Efficiency* refers to the fulfillment of objectives with minimal use of resources. These resources may take the form of time, space, or material inputs, depending on the domain of application.

In the realm of engineering, a common resource is that of energy. This factor may be found in two guises: the cumulative energy consumed, or its rate of use, also known as *power*.

In the realm of business and commerce, most inputs and outputs are expressed in terms of the common denominator of money. For this reason, each factor is assigned a financial value which contributes to the overall cost or benefit of a system.[3] In many situations, time must also be considered as a resource to be economized.[4]

Considerations of efficiency for any purposive system must be subject to the fulfillment of functional requirements. There is little point in designing a system to be produced or operated at zero cost, if it does not serve any function.

## Relationship to Effectiveness

The quest for efficiency must be subject to the fulfillment of the objectives for an engineered system. Such a strategy may be reflected in both hardware and software. An example of a hardware approach is found in the substitution of cheaper materials, such as the use of copper rather than gold to serve as an electric conduit even at the expense of decreased functional performance. A structural example is found in the nestling of a house into a hillside to conserve energy, or the stamping of holes in metallic struts to conserve material and reduce weight.

The use of enzymes to catalyze chemical reactions illustrates the efficient deployment of resources in biological systems. These enzymes accelerate physiological processes, whether in the synthesis of new products or in catabolism such as digestion. Like marriage brokers, the enzymes bring together unattached molecules that bind to special receptor sites. These molecules then join in harmony or interfere destructively according to their innate dispositions. Without such catalysts, the corresponding activities would proceed with reasonable speed only at higher temperatures. These elevated temperatures would obviously call for gluttonous rates of energy consumption. Further, life as we know it could hardly be possible if such searing temperatures were required. As a side effect, the attendant heat would cause structural disruptions in neighboring organs by severing their organic chemical bonds.

Even if all the organs of the body could withstand the heat, thermodynamic principles would render life impossible. The second law of thermodynamics states that a system cannot produce more useful energy than it takes in. Hence the energy cost of keeping the body in operation despite heat loss through convection, conduction, and radiation—in addition to the kinetic energy needed for locomotion in procuring food—would likely exceed the energy value from the ingested food.

The use of catalysts such as enzymes allows for the appropriate chemical reactions to proceed at a reasonably rapid pace. To the extent that the most effective chemical pathways may be supported by catalysts, this mechanism represents an optimizing strategy.

Ongoing scientific advances make feasible systems of increasing efficiency. A nanocomputer—a computing device of molecular scale in which each component is made up of a few atoms or molecules—is an example of such foreseeable progress. Current advances in biotechnology, chemistry, and micromanipulation technology point the way toward these computers of atomic dimensions. Such devices will be characterized by extreme efficiency in space and energy consumption, in addition to high computing speeds.[5]

In the software arena, efficiency considerations are paramount in computational procedures. In this domain, great effort is expended to avoid the cost of an exhaustive search of all the potential solutions. To this end, various pruning techniques are used in searching a tree of alternatives for a solution. Strategies for efficiency may lead to the optimal solution or they may not. Hence we may categorize such strategies into two classes:

- *Optimizing*. An optimizing strategy is one that is guaranteed to yield the best solution to a given problem.
- *Satisficing*. A satisficing strategy is one that leads to a satisfactory solution. The result may not reflect the optimum, but an adequate solution.

An example of an optimizing strategy from mechanical design is: "If a structure is designed to carry a load, ensure that all its substructures which are in series are of similar carrying capacity." This is a domain-specific example of the maxim "A chain is as weak as its weakest link." The concept occurs in other fields such as operations research, where it is known as the Minimum-cut/Maximum-flow principle: the maximum flow of objects from one end of a network to the other is defined by the minimum value of the flows through all possible transverse partitions (cuts) across the network.

A satisficing strategy is appropriate when no optimizing strategies are known, or they are known but are too inefficient to implement. An example of a satisficing strategy from engineering design is, "Relax tolerance requirements when functional objectives are not impaired." An example from civil engineering is, "When human life is at stake, as for a bridge or a building, put in a safety factor of 20." A 20-fold expansion over the minimal level of physical strength required for functional fulfillment is a convenient rule of thumb to balance human safety and production cost.

An example of a satisficing heuristic from the realm of factory design is: "Locate the shipping and receiving departments closest to the main roadway." This rule of thumb may not always yield the best results, but usually leads to a good solution.

A heuristic from the marketing field may be: "Five percent of corporate revenues should be spent on advertising." A rule such as this says nothing about the modes of expenditure, hence can only serve as an approximate guideline at best.

An example from knowledge engineering is the cut-off strategy used in conjunction with decision rules of questionable validity: "If a line of reasoning leads to an intermediate conclusion whose validity appears tentative, discard that train of thought." This rule may result in computational efficiency by allowing other, more promising avenues to be pursued. But it may also lead to suboptimal results, since the intermediate conclusions—tentative as they are—might have led to correct results.

The relationship between efficiency and performance is depicted for each class of strategies in Figure 8.1. The most straightforward approach to a design problem is linear search: try all combinations in sequence until the best one is found. Unfortunately, this approach is not feasible when the search space is infinite, such as when continuous variables are involved. Moreover, even when the search space is finite, the procedure may be infeasible in practical terms.

For most straightforward strategies for enhancing computational speed, the "goodness" of the resulting solution decreases as the efficiency of the alternative decreases. A heuristic in point is the simplistic strategy "Identify and evaluate every fifth candidate solution." Such an approach is more likely to yield suboptimal results if and when an adequate solution is found.

## Efficiency versus Reliability

The effectiveness of a physical system depends on its absolute levels of performance under routine operating conditions. However, any realistic system is susceptible to intermittent or complete failure, thereby affecting the amount of time available for effective performance. The availability of a system is, in turn, the complement of its reliability.

In general, reliability may be enhanced by deploying more resources and there-

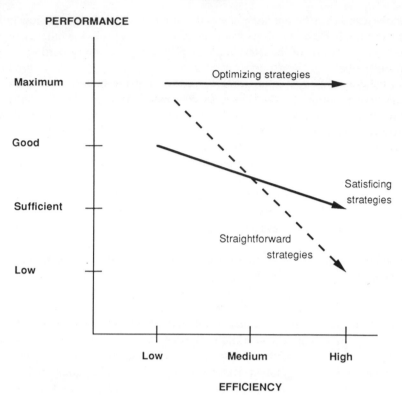

**Figure 8.1** Efficiency versus performance for various classes of strategies.

by sacrificing efficiency by increasing the cost at some stage of the product life cycle, whether in fabrication, operation, or maintenance. The system architect can increase reliability by either increasing the overall reliability of each component, or by creating redundancy. For example, a mechanical component may be supported by a backup to be pressed into service if the primary mechanism fails.

In the biological realm, the hormone luteotropin is a good example of efficiency due to its versatility. It is synthesized in the pituitary gland in response to signals released by the hypothalamus. Commonly known as luteinizing hormone in females and as interstitial cell-stimulating hormone in males, this substance has different functions in each gender. In females, luteinizing hormone is involved in preparing the mammary glands for secretion and in forming the corpus luteum in the ovary. In males, it controls the secretion of interstitial cells in the testes. Thus, not only does the hormone have different functions in males and females, but it is also involved in distinct tasks at different times in the female menstrual cycle.

The structures supplying blood to the brain contain a redundant feature to safeguard against the likelihood of systemic failure. The brain obtains its blood supply from two main sources, the vertebral and internal carotid arteries. The internal carotid arteries, one located on each side of the head and neck, normally carry 85 percent of the blood supply, while the vertebral arteries convey the remaining 15 percent. These two vessels enter the brain space at the base of the skull and are connected within the brain itself by two other lateral blood vessels: the anterior and posterior communicating arteries. The resulting "vessel" forms a loop, known as the circle of Willis. Additional vessels convey the blood from this loop to all parts of the brain. The redundancy embodied in the vascular loop, ensures that the brain receives nourishment even if one of the blood vessels is blocked.[6]

In the neurological domain we find an example of efficiency through physical divergence. After emerging from the cell body of the neuron, the axon often splits into several branches which then synapse with other neurons. This type of divergence enables the simultaneous transmission of information (afferent or efferent, depending on the original neuron) to many different sections of the central nervous system. In this way the output from a single action potential can be dispersed, having a considerably enhanced effectiveness. Due to divergence, each motor axon can innervate many muscle fibers, ranging anywhere from 15 in the eye to 1,900 in limbs.[7]

The two adrenal glands in the human body illustrate the concept of economy through physical integration. Each adrenal gland consists of two components: the inner adrenal medulla, and the adrenal cortex that surrounds it. The medulla produces the hormones epinephrine and norepinephrine, substances that enhance the activities of the sympathetic nervous system. The adrenal cortex, on the other hand, secretes several other hormones. The outermost layer of cells in the cortex synthesizes aldosterone, which helps maintain sodium and potassium balances. The rest of the cortex produces both cortisol, which regulates carbohydrate and protein metabolism, and the sex steroids such as testosterone. Physical integration allows for the sharing of structural elements that support more than one process or function.

A perennial issue in the design of any product and its fabrication process is the trade-off between cost and quality. If cost is viewed as a measure of the efficiency in the use of the factors of production, then decreased cost is a laudable objective. However, reliability may be sacrificed in the quest for efficiency. To illustrate, relaxing the tolerance requirements on production operations for a motor may decrease short-term fabrication costs while maintaining minimal quality constraints. However, the loose tolerances may result in increased vibrations, accelerated wear, and premature failure during operation. As a result, the long-term cost in terms of maintenance and loss of customer goodwill may exceed any initial cost advantage.

## Summary

Efficiency relates to the satisfaction of objectives with minimal use of resources. These resources may take the form of time, space, or material inputs, depending on the application domain.

- Design rules for enhancing efficiency may be of the optimizing or satisficing variety. The former type results in the best possible solution given the problem description, while the latter type yields a solution that is "good enough."
- The effectiveness of a system is defined by its performance levels under normal operating conditions as well as its availability.

- The availability of a system depends on its reliability. These two parameters may often be enhanced at the cost of efficiency.
- Efficiency must be considered for the initial fabrication of a system as well as its long-run operation. In this way, the initial cost disadvantage involved in enhancing the quality of a system may be outweighed by enhanced utility over the entire system life cycle.

# III

# INTERFACTOR
# TRADE-OFFS

*The same capital will in any country put into motion a greater or smaller quantity of productive labour, and add a greater or smaller value to the annual produce of its land and labour, according to the different proportions in which it is employed.*[1]

Adam Smith

*The decisive properties of computing machines involve balance: balances between the speeds of various parts, balances between the speed of one part and the sizes of other parts, even balances between the speed ratio of two parts and the sizes of other parts.*[2]

John von Neumann

# 9

## Space versus Time

*In the pre-relativity physics, space and time were separate entities . . . . That there is no objective rational division of the four-dimensional time continuum into a three-dimensional space and a one-dimensional time continuum, indicates that the laws of nature will assume a form which is logically most satisfactory when expressed as laws in the four-dimensional space-time continuum.* [1]

Albert Einstein

The trade-off between time and space is a ubiquitous consideration for the design of intelligent systems. In this domain, as elsewhere, a constraint on either time or space results in increased demands on the other variable. If the space available for housing the system is very limited, the system will need more time to perform its various tasks. However, if a great deal of space is available for the system, then it can function more easily and rapidly.

In the neurological sphere, the two types of impulse summation occurring in neurons illustrate this time-versus-space trade-off. Since a single excitatory stimulus cannot depolarize a post-synaptic neuron sufficiently to trigger a fresh impulse, only the collective effect of many synapses can elicit a response from the cell.

Summation occurs in the post-synaptic neuron either temporally or spatially. Temporal summation refers to the ability of one pre-synaptic neuron to transmit several excitatory impulses over a short period of time, which the post-synaptic neuron then sums. In contrast, spatial summation denotes the simultaneous impulse transmission across two or more synapses at different places on the post-synaptic neuron.

The temporal summation of signals allows for the use of a small area for receptors to receive the initial stimulus. But an operating constraint on the rate of signal processing is the length of the refractory period for regenerating the electric potential. In contrast, although spatial summation is not hampered by the refractory period between axonal transmissions, it requires a larger area for receiving the initial stimulus, so that two or more inputs can share the burden of triggering an excitatory potential.

The human body involves many processes for which time is of the essence. A number of organs feature large surface area to expedite such processes as absorption. The organ responsible for gustation, the taste bud, is a good example of this strategy. There are approximately 10,000 of these organs in humans, located on the papillae of the tongue. A taste bud is a composite structure consisting of a dozen or so receptor cells. Each receptor cell possesses hair-like projections, or microvilli, which extend from the cell body into the fluid-filled environment. Chemicals enter the cells and stimulate receptors on the microvilli. The absorption of substances changes the membrane potential of the sensory cell and generates a neural impulse. Once initiated, the neural signal travels to the brain stem for further processing.[2]

The location of taste receptors on microvilli greatly increases the surface area available for gustation. The increase in space corresponds to a decrease in the time required for the detection of one of the four primary qualities: sweet, salty, sour, and bitter. In this way, microvilli embody the trade-off between time and space.

The same principles relate to absorptive processes in the small intestine and kidney. Although the coils of the small intestine would measure several feet if extended, digested food travels rapidly through the intestine, limiting the amount of time available for absorbing nutrients. To maximize absorption while operating under this time constraint, the small intestine possesses many small villi. In a similar way, waste products from digestive and cellular processes passing through the kidney allow for water to be reabsorbed before the waste products are excreted. Since the objective is to remove as much water as possible within the limited time constraint, the renal system has developed a network of coiled tubules through which the waste products must pass. The kidney and intestines further illustrate the spatiotemporal connection.

Time and space are often tradeable resources where engineered objects are concerned. When space is at a premium in a factory, for example, production operations can be hampered by the restricted movements of machine tools as well as the convoluted flow of works-in-process and limitations in inventory levels. The end result may be an unacceptably high value of throughput time for workpieces. In this situation, throughput time may be decreased by utilizing more space, and vice versa.

The relationship between space and time is a well-known problem to students of computer science. The field of complexity theory deals with resource allocation requirements for computer algorithms; a major aspect of this field is the study of the trade-offs between the time and memory required by alternative procedures that accomplish equivalent tasks.

One example occurs in the process of sorting records in a large database whose size exceeds the memory capacity of the computer. This would be analogous to using the top of a file cabinet rather than a large desk for collating copies of a manuscript. Without even delving into specifics, it is apparent that a sorting procedure for the constrained situation will be more involved—and take more time— than one in which the memory size is sufficient to accommodate the entire database at all times.

## Summary

The dimensions of intelligent systems are independent in the sense that one cannot be regarded as a simple algebraic composition of the others. On the other hand, the parameters relating to disparate dimensions are dependent to the extent that a change in one can adversely affect the status of another, or one can be substituted in part for the other. An important example of interdimensional trade-offs lies in temporal and spatial resources. These are often complementary factors that reinforce each other to fulfill the objectives of an intelligent system.

- As partially tradeable resources, time and space can often be substituted for each other while maintaining a fixed level of system performance.
- The finite speed of reactions through diffusion and chemical pathways leads to diverse examples of process acceleration through increased surface area or volume in the biological realm.
- Artificial systems can also be designed to take advantage of the space-time trade-off.

# 10

## Mechanism versus Process

*Mind and world in short have been evolved together, and in conse-quence are something of a mutual fit.*[1]

William James

The structure of the machine or of the organism is an index of the performance that may be expected from it. *The fact that the mechanical rigidity of the insect is such as to limit its intelligence while the mechanical fluidity of the human being provides for his almost indefinite intellectual expansion is highly relevant . . . if we could build a machine whose mechanical structure duplicated human physiology, then we could have a machine whose intellectual capabilities would duplicate those of human beings.*[2]

Norbert Wiener

The fulfillment of a goal requires coordination among a sequence of purposive events and the mechanisms for their realization. At times the mechanisms complement the processes through direct support, and at other times through substitution or even redundancy.

### Hardware versus Software

Hardware and software relate to the partitioning of function among physical and procedural elements. Hardware refers to the physical aspects of implementing the desired function. Software, on the other hand, refers to the programming required or the relationships among the system components.

The movement of food in the swallowing process is a biological function shared by both the hardware and software of the human digestive system. As food is chewed in the mouth and begins movement through the digestive tract, it travels first from the mouth to the esophagus. In so doing, the mass of chewed food must avoid going up through the nasal passage or down through the trachea. As the bolus of food moves to the rear of the mouth, it elevates the soft palate—the roof of the mouth—and thus mechanically seals off the passage to the nasal cavity. In this way, food is effectively prevented from entering the nasal passage by accident.

At the same time that the bolus is forcing the soft palate upward, pressure receptors are stimulated in the pharynx at the rear of mouth. The receptors then send impulses to the swallowing center in the medulla. The swallowing center generates commands that inhibit respiration while causing the pharyngeal walls to move upward and inward, and thereby providing a clear path for the bolus. This, in turn, closes the glottis, preventing food from accidentally entering the trachea. In this way both hardware and software mechanisms cooperate to implement the swallowing function.[3]

The hardware of the human circulatory system maintains the flow of blood in specific directions. An example is found in heart valves, a mechanism activated by the very agent it is designed to control.

In the heart, blood first flows into a chamber known as the right atrium and passes through the atrioventricular valve into the right ventricle. From here, blood moves through the pulmonary valve into the arteries and capillaries of the lungs, where carbon dioxide is exchanged for fresh oxygen. The blood next enters the left atrium of the heart, followed by a passage through a second atrioventricular valve into the left ventricle. From this chamber the blood flows past the aortic valve into different parts of the body to provide oxygen and nutrients to the limbs, brain, and other organs.

The difference in pressure across heart valves prompts their opening and closing, thus regulating both the amount of blood flow and its direction. For example, the atrioventricular valve permits blood flow only from the right atrium to the right ventricle. During this passage, the blood pushes away the leaves of the valves to create an opening. When the ventricle contracts, the increased pressure of the blood in the chamber forces together the leaves of the valve, thereby sealing off the orifice.

The leaves of each valve are attached to the ventricular wall by fibrous strands. These attachments prevent the leaves from retreating into the atrium. In this way, the unidirectional flow of blood is determined by the hardware of the heart.

The hardware of the human sensory organs supports the reduction of the large amounts of data impinging on these systems. Between the time that incident light first reaches the eye and activates the photoreceptors, and the arrival of electrical signals in the visual cortex, a great deal of data reduction occurs in the hardware of this information-processing system. This reduction of information actually takes place in the rods and cones, as well as the neuronal circuitry leading from the eye to the visual cortex.

A case in which both hardware and software models have been used to accomplish the same task relates to the cash register. Before the advent of computer chips, cash registers were purely mechanical in construction, yet they were able to implement logic in performing such tasks as addition and subtraction. The information-processing capabilities were superimposed on a mechanical apparatus. In contrast, modern electronic cash registers process information in a similar fashion, but use software to implement the arithmetic procedures.

## Storage versus Computation

The contrast between computation and storage exemplifies the opposition between
process and mechanism, respectively. Storage refers to the retrieval of an item of
knowledge directly from a repository of knowledge rather than through a process of
derivation. In contrast, computation refers to the generation of results through
transformations on an initial set of data or knowledge. Even in computation, however, the initial knowledge base must be obtained from some external source such as
an intelligent agent or archival storage.

To illustrate: suppose that the mass, volume, and density of a particular product
are of interest to a designer. The values of all three properties may be stored in a
database; or any two (e.g., mass and volume) may be retained and the third (density) calculated when needed. The first scheme corresponds to the storage approach,
while the second scheme represents computation. In the production arena examples
of storage versus computation are found in the sale of prefabricated suits versus
tailored suits, or a mass-production factory versus a special-order job shop.

The human brain is likewise capable of computing or storing information,
depending on the mode of input or frequency of access. If an employee is asked
what her salary is per minute, she will calculate it and arrive at an answer. If,
however, several people ask her the same question within a short time interval, the
response is likely to be memorized. The first case reflects a computational approach, while the latter situation relies on storage.

According to one adage, experience is what you have when you don't need it
anymore. This aphorism embodies the view of human memory as a mechanism for
storing knowledge of past events, and thereby obtaining the benefit of historical
results without fully repeating the original sequence of actions.

In the human nervous system, neurons rely on a combination of storage and
computation to retrieve needed data. The output of a neural signal is computed as a
function of its input signals. This output may be of two types: graded potentials and
action potentials. Graded potentials are local changes in membrane potential; they
are usually generated in response to changes in a neuron's environment. The magnitude of the change in membrane potential varies according to the magnitude of the
environmental change. Hence, the response is computed as a function of the stimulus intensity.

In contrast, an action potential refers to the output of a fixed voltage signal
induced by an input signal; the input signal may have any value above a given
threshold. The depolarization of a membrane changes its potential from $-70$ to $+30$
millivolts regardless of the exact value of the input signal. This voltage inversion is
caused by the movement of sodium and potassium ions across the cellular boundary
in response to changes in membrane permeability to these ions. The amplitude of
the action potential can therefore be viewed as the result of information encoded in
the hardware structure of the axon and replayed on command rather than computed
from input signals.

Each neuron in the nervous system determines whether an impulse stimulus is
strong enough to induce its firing. This is achieved by computing the total sum of
the positive and negative currents that are induced in the "recipient" neuron's

dendrites and cell body. Neurotransmitters released by the axon of the pre-synaptic neuron induce these currents. However, the computation performed by the post-synaptic neuron is *not* merely an arithmetic sum. Voltage in the post-synaptic neuron is influenced by the strength of its synapses, the arrival time of different inputs, and the relative position of synapses on dendrites and the cell body of the neuron.[4] In this way, the computed result depends in part on the storage of information over brief time intervals.

In genetics the coupling of computation and storage efficiently serves the system's informational needs. Within an organism, each cell contains the same genes stored on the same number of chromosomes. But the activation of specific genes follows from some processing or computation of information derived from the cell's environment. Thus, the production of the organism and its everyday operations results from a combination of computation and storage. This combination maximizes the efficiency of the system, since the information necessary for the proper response is always available, but the degree of response can be tailored to the particular situation.

For a diverse array of body functions, the information that associates specific responses with the ability to execute them is encoded within some memory system. However, the degree of response is usually determined through computation. For example, the genetic information needed to synthesize hormones is stored, but the amount of production is continually computed. Similarly, the body has certain responses for decreasing elevated body temperature, such as vasodilation and perspiration. However, the degree to which the vasodilation and perspiration occur varies over time and must therefore be computed. Again, this combination enables the system to function efficiently by yielding the proper level of response to each situation.

In algebraic applications, trigonometric functions such as the sine and cosine are often calculated in computer systems. Although retrieval of such facts from memory might take less time, the storage of all of these values would require far too much space. Thus, the software is designed to calculate these functions.

Parallel processing strategies—whether in computer algorithms, the factory environment, or corporate departments—may also be viewed as a way to reduce the time required for a task by utilizing more space. As discussed in Part II, however, parallel strategies entail greater overhead in terms of coordination requirements. In this case efficiency obviously suffers. In addition, the extra overhead may be so onerous that the overall task time may even increase!

Since the computational approach must begin with some initial knowledge base, it is possible to regard this method as a generalization of the storage approach. More specifically, storage is the limiting case of computation in which the null transformation is applied to some knowledge retrieved from memory.

## Summary

The architect of a system must decide whether the bulk of the information processing will require computation or the storage and recall of many facts. Since any system will have a finite amount of space available for the equipment to compute and store details, a trade-off exists between computation and memory.

- Mechanism and process are complementary factors. They support each other and are partially substitutable.
- Software is the representation of knowledge or procedures encoded in a programming language. It is used to define a process for obtaining desired outputs in response to some input.
- Storage refers to the representation of knowledge in static form on some physical device.
- Computation denotes the calculation of a result through a set of procedures or knowledge encoded in software.

# IV

# APPLICATIONS

*Five areas of management constitute the essence of proactive performance in our chaotic world: (1) an obsession with responsiveness to customers, (2) constant innovation in all areas of the firm, (3) partnership . . . with all people connected with the organization, (4) leadership that loves change (instead of fighting it) and instills and shares an inspiring vision, and (5) control by means of simple support systems aimed at measuring the "right stuff" for today's environment.*[1]

Tom Peters

# 11

## Autonomous Robot

*There is no right to deny freedom to any object with a mind advanced
enough to grasp the concept and desire the state.*[1]

Isaac Asimov

People have longed for many things over the millennia. One of these is the creation
of a twin, both in form and function. We are finally on the verge of realizing this
dream as a result of advances in genetic engineering, artificial intelligence, materials science, silicon technology, and mechanical engineering. At the current state
of technology, there appear to be two equally promising avenues to the destination—organic and inorganic routes. The first route will depend on molecular engineering, following a translation of the book of life encoded in the deoxyribonucleic
acid (DNA) of each human cell. In creating organic intelligence, it is not necessary
to translate the entire DNA sequence, merely enough to regulate the generation of
organs to trigger the growth of an entire animal.

The inorganic route to intelligence is currently based on silicon technology. The
final product will rely on the confluence of miniaturized circuitry, parallel processors, materials technologies, and mechanical arts.

In this chapter we explore the critical issues in designing a mechanical robot.
The machine will have modest capabilities in terms of running errands and conduct-
ing conversations. This type of robot will serve as a precursor to the android, a
machine characterized by human-like structures such as supple skin. The large-scale
development issues, however, will not likely vary much among the autonomous
robots, ranging from the basic rover to the humaniform agent.

## Purpose

Autonomous robots may be classified, among other parameters, in terms of their
similitude to humans. Figure 11.1 presents a typology of automated devices ranging
from simple mechanisms to life-like machines. An *automaton* is any machine
capable of independent operation, such as a clothes dryer. A *flexible machine* is a
special case of an automaton with versatile capabilities; it possesses a range of
behaviors that can be programmed as the need arises. An example is a welding robot
on the factory floor that can be configured to other production operations. A *mobile*

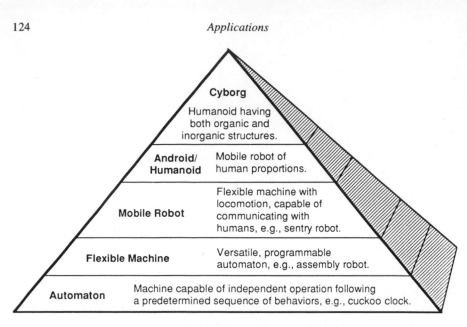

**Figure 11.1** A typology of intelligent machines.

*robot* is a flexible machine capable of roaming its environment. It can partially select its own goals and communicates with other agents, including humans. An *android* or *humanoid* is a mobile robot whose structure roughly resembles that of humans. Finally, a *cyborg* is a humanoid incorporating organic structures and exhibiting some physiological processes similar to those of humans. In this chapter we focus on the design of mobile robots and androids.

Our goal is to explore the issues in designing a general-purpose robot that can conduct a basic conversation with its master and assist in household duties such as cleaning and watching over children. The objectives of the robot are as follows:

1. Perform simple errands, such as fetching a cup of coffee or depositing a letter in a mailbox.
2. Conduct conversation, including providing information in response to factual questions.

In an ordinary household, the robot may be used to clean rugs, set tables, and wash dishes. In a factory setting, a variant model of the machine might be used to fetch tools, convey messages, or assist in production activities.

The second objective indicates that the robot must be able to engage in conversation and assist in resolving problems that require logical deduction. This ability includes the capability for providing factual answers to questions. To this end, the robot can incorporate knowledge from conventional sources such as dictionaries, encyclopedias, handbooks, and other references. If the robot is equipped with an internal radio and telephone, it can access broadcast media as well as electronic database services, then sift through the data to formulate an intelligent response.

The robot, as a product, will be subject to a number of constraints. The primary constraints on the robot are as follows:

1. The robot must require no special equipment other than that available in an average home. For example, it must be able to walk up a flight of stairs rather than require an elevator.
2. It must cost no more than $10,000, slightly under the average cost of an automobile in 1990.

The specific functional requirements will depend on the ultimate robotic application, whether as a maid, an errand runner, factory hand, or some other role.

# Space

## *Environment and Technology*

The human aspiration to create animated objects probably germinated with the first twinklings of humanity. Since the Renaissance, increasing numbers of artifacts have been created to perform sequential tasks ranging from playing musical instruments to drawing figures and writing script.

In 1922, the Czech playwright Karel Capek introduced the word robot to the world (see Table 11.1). The term comes from Czech words meaning "work" and "serf." In the play *R.U.R. (Rossum's Universal Robots)*, Capek portrayed a future in which the robots become so intelligent and disenchanted with humans that they revolt, destroy their masters, and seed a new world of robotic heritage.[2] This play was one of many instances in science fiction that portrayed robots as ingrates rising against their creators.

The benign view of robots living in harmony with humans was popularized by the science writer Isaac Asimov. He coined the term *robotics* as a field of study and wrote a series of stories begining in the early 1940's.[3] These stories focused on the "Three Laws of Robotics" that govern the behavior of sentient machines:

1. A robot must never harm a human being, or, through inaction, allow a human being to come to harm.
2. A robot must always obey a human being, unless this is in conflict with the first law.
3. A robot must protect itself from harm, unless this is in conflict with either of the first two laws.

Asimov explored the consequences of these regulations and the dilemmas they engender: for instance, should a robot tell the truth and bring psychological harm to a particular human, or tell a lie and spare her the anguish? Given the rapid advances in robotic capability and concern about the future welfare of humanity, the three laws or some variant thereof may well become an ultimate guiding force in robots.

In 1954 the American inventor George Devol began patent work that would lead to the industrial robot. Two years later, he teamed up with Joseph Engleberger, an engineer, to form the Unimation Company. The firm developed flexible industrial

*Table 11.1* Historical highlights in robotics

| Year | Event |
| --- | --- |
| 1921 | Czech playwright Karel Capek introduces play on London stage: *R.U.R.* (*Rossum's Universal Robots*) |
| 1954 | George Devol develops first programmable robot |
| 1954–63 | First generation robots |
| 1967 | Quadruped personnel carrier at General Electric |
| 1967–69 | Shakey: mobile robot at Stanford Research Institute |
| 1973 | Cart: mobile robot at Stanford University |
| 1980 | Perambulating Vehicle: successful quadruped at Tokyo Institute of Technology |
| 1984 | One-legged hopping machine at Carnegie-Mellon University |

machines and began to market them in the early sixties. Since then, a myriad of companies have entered the robotic market.

Some of the most popular arenas for industrial robots have been in materials handling, including the transportation of workpieces within the factory and their set-up on machine tools. The popular production operations have ranged from plastic molding and die casting to stamping and welding.

The productivity of industrial robots springs from their consistency and dedication rather than sheer speed. For example, a welding robot can form a seam no faster than a human worker. On the other hand, the robot does not tire and therefore can keep the torch on the workpiece 80 percent of the time rather than a human's 30 percent.[4] A robot can also work almost four shifts per week, including weekends, with the exclusion of downtime for maintenance activities. Further, consistency in production operations implies higher quality and lower rejection rates in the workpieces.

Certain production activities have been resistant to automation. A case in point is the handling of flexible materials such as fabric, which may fold over itself and thereby confuse sensing as well as handling devices. Another example is the task of picking up objects of irregular shape when they are presented collectively in haphazard fashion in a bin. The profusion of lines and shadows wreaks havoc in the visual recognition apparatus. In fact, many existing vision systems work effectively only in structured environments where objects are presented in highly constrained positions, while distracting reflections and shadows are held to a minimum.

The 1960's heralded systematic research in mobile robots. Between 1967 and 1969, investigators at the Stanford Research Institute developed a wheeled robot named Shakey.[5] Sensory capabilities for this machine included bump detectors, a sonar range finder, and a television camera for visual interpretation of the environment. Shakey could navigate around obstacles and roam the hallways. The machine represented a remarkable achievement. However, it was considered a failure at the time because of its reliance on a separate mainframe computer. The computer handled much of the reasoning processes and conveyed its commands to the robot through a radio channel.

In 1973 development work began on the Cart, a roving machine built by Han Moravec and others at Stanford University.[6] A distinguishing feature of the machine was a television camera that could move laterally along the breadth of the cart, thereby allowing for three-dimensional imaging through multiple perspectives. Major liabilities of the Cart included the lack of a closed-loop distance encoder and laborious computational requirements. It could move only about a meter at a time; at each stop it would wait for multiple images from the television camera, interpret the scene, and plan its next move.

An important feature of general-purpose robots is the flexibility born of legged locomotion. In 1967, the General Electric Corporation developed a four-legged machine for the Department of Defense, for use in environments inhospitable to wheeled or treaded devices.[7] The machine carried a human operator who had to control each of the four legs. This proved to be an inordinate demand on the driver, and the machine regularly tipped over. Although the project failed, it served to inspire more successful efforts.

One such device was a quadruped developed at the Tokyo Institute of Technology in 1980. The Perambulating Vehicle/Mark II was a four-legged device that relied on a layered decomposition of locomotive control. At the top level, an intelligent controller or human would decide on a walking strategy in terms of direction and speed. The controller relied on a vision sensor at the second layer. Information about the terrain was used to generate the gait commands, including the foot placements needed to ensure smooth motion. The third layer maintained stability by relying on tactile detectors embedded in the feet as well as posture sensors in the frame of the machine. The lowest layer involved the transmission of commands to individual servomotors driving each joint.

In 1983, a six-legged robot was developed by Odetics, Inc. for commercial production. A battery-powered model, Odex I, relied on a radio channel for leg control and a video link for conveying images. The machine could walk over obstacles and lift loads weighing several times its own weight.

Research continues on machines that rely on one or two legs. In the early 1980's Marc Raibert developed one-legged hopping robots at Carnegie-Mellon University.[8] This line of research is being continued at the Massachusetts Institute of Technology, where two-legged machines cavort on the laboratory floor.[9]

Prototypes such as these serve as vehicles for enhanced understanding of legged locomotion. Depending on the environment, legs might not be as efficient or effective as other means of conveyance. Even so, their flexibility and similarity to our own form will ensure their implementation in many robots of the future.

## Internal Requirements

A robot must be able to cope intelligently with its environment, both during routine operation and when faced with unexpected events. The interfaces involve input mechanisms such as sensory devices as well as output apparatus such as manipulators and communication devices.

A mobile robot must pack a remarkable amount of hardware and software capabilities into a volume no larger than that of a person. These include physical components such as energy storage devices and actuators. The machine must also process large quantities of information in parallel.

Supercomputers in the late 1980's draw on a memory of roughly 100 billion binary digits. These machines can perform one billion operations per second on chunks of information 64 bits long; the processing capacity therefore approaches a hundred billion bits per second. These numbers are comparable to the neural sophistication of rodents. In contrast, the human neural system has a memory capacity in the neighborhood of a quadrillion bits of information, and a processing capacity of 100 trillion bits per second.[10] To match the performance of the human brain, the capacity of computational hardware must increase by a factor of ten thousand in memory and one thousand in speed. This magnitude of computational power is not likely to become available in the size of a breadbox until the first decades of the new millennium.

The design goals for our autonomous robot, however, are much more modest. We do not seek a match for humans in all spheres of mentation and physical prowess, only in deductive reasoning and basic ambulation. The powers of reasoning that we hold so dear and which differentiate us from our mammalian cousins developed only in the last few million years of evolution, as opposed to the billions required to acquire basic capabilities in locomotion, animation, proprioception, and reproduction.

From this perspective, it is not surprising that high-level reasoning processes are the easiest to translate into software. Since the 1960's, a number of smart programs have matched, and sometimes exceeded, the performance of humans in fields ranging from mathematics to chemistry, and engineering to finance. Our ability to construct miniature physical devices and intelligent software is sufficient to support the development of autonomous robots by the turn of the century.

Table 11.2 depicts the development of robotic technologies over the next few decades. The table indicates, for example, that vision is still an infant technology. Commercial vision systems before 1990 could deal only with structured environments such as an assembly line equipped with special lighting. Moreover, the vision modules have problems identifying jumbled collections of objects, such as pieces of fabric in a bin. The recognition and identification of objects in the real world are being demonstrated in laboratory prototypes such as autonomous vans. The technology will likely find its first commercial uses in the 1990's and come into widespread acceptance at the beginning of the next millennium.

## Structure

The structure of an autonomous robot will be defined by its particular functional requirements. For example, an automaton whose sole purpose is to vacuum floors need not differ much from that of a traditional vacuum cleaner with the exeption of sensors and smarts in lieu of a handle. On the other hand, a general-purpose robot capable of dusting, cooking, baby sitting, and other functions will likely take a more anthropomorphic shape.

*Table 11.2* The maturation of some robotic technologies

| CAPABILITY | Pre-1990 | 1990's | Post-2000 |
|---|---|---|---|
| **Manipulation** | | | |
| Industrial Tools | ▬▬▬▬ | | |
| Anthropomorphic | ⅠⅠⅠⅠⅠⅠⅠ▬▬▬▬▬ | | |
| **Communication** | | | |
| One-way (orders to robot) | ▬▬▬▬ | | |
| Bilateral | ░░░░ⅠⅠⅠⅠⅠⅠⅠ▬▬▬ | | |
| **Locomotion** | | | |
| Rails | ▬▬▬▬ | | |
| Wheels | ▬▬▬▬ | | |
| Tracks | ▬▬▬▬ | | |
| Legs (biped) | ░░░░ⅠⅠⅠⅠⅠⅠⅠ▬▬▬ | | |
| **Sensing** | | | |
| Smell (limited odors) | ▬▬▬▬ | | |
| Vision (real-world) | ░░░░ⅠⅠⅠⅠⅠⅠⅠ▬▬ | | |
| Voice (real world) | ░░░░ⅠⅠⅠⅠⅠⅠⅠ▬▬ | | |
| Touch (texture, etc.) | ░░░░ⅠⅠⅠⅠⅠⅠⅠ▬▬ | | |
| **Reasoning** | | | |
| Isolated Data | ▬▬▬▬ | | |
| Database | ⅠⅠⅠⅠⅠⅠⅠ▬▬▬▬ | | |
| Knowledge Base | ░░░░ⅠⅠⅠⅠⅠⅠⅠ▬▬ | | |
| Self-Learning | ░░░░ⅠⅠⅠⅠⅠⅠⅠ▬▬ | | |

| | | |
|---|---|---|
| ░░░░░ | = | Laboratory prototypes |
| ⅠⅠⅠⅠⅠⅠⅠ | = | First commercial applications |
| ▬▬▬▬ | = | Widespread commercial applications |

## *Major Components*

The functional subsystems of a robot are presented in Figure 11.2. In addition to the physical body which accommodates all other subsystems, the hardware components may be classified into the sensors and effectors. The software aspects involve the interpretation of sensor readings into conceptual objects that may be used by the high-level reasoning component. If the robot is to respond in some way to the

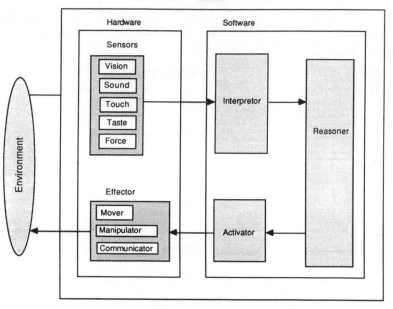

**Figure 11.2** Major components of a robot.

environment, the results of the reasoning process must be translated by the activator component into low-level commands. These commands drive the hardware effectors, whether in terms of communicating with external agents, manipulating objects in the environment, or moving the robot to new locations.

The hardware components are explored further in this section. The software modules are examined in a subsequent section in the context of processes required to implement intelligent activities.

## *Reasoning Module*

As with the human nervous system, the complexity of the reasoning module can be addressed by a layered decomposition of functions. One such approach consists of several levels ranging from collision avoidance to reasoning about behaviors:[11]

- *Avoid objects.* The robot must avoid colliding with other objects in the environment.
- *Wander and explore.* A robot should be able to roam and head for destinations that seem interesting and within reach.
- *Build maps.* The explorations should provide information for building a map of the locale and for planning paths between places.
- *Monitor changes.* The reasoner should notice changes in the external world.
- *Identify objects.* The robot should recognize items in the environment and be aware of the potential actions that may relate to them.

- *Plan external changes.* The reasoner should pursue its objectives by creating plans. In general, the implementation of these plans results in changes to the world.
- *Reason about external behaviors.* The reasoner must be aware of the properties and activities of external objects and upgrade its plans accordingly.

Each level of competence subsumes the capabilities in all the lower layers. A given layer of control guides the activity of its subordinate levels; on the other hand, a particular layer operates without explicit knowledge of its superior level, except to the extent that the higher level constrains its own space of behaviors.

## *Locomotion*

An autonomous robot, almost by definition, must have the capacity to move about from one locale to another. The most common modes of locomotion, as well as their respective advantages and drawbacks, are given in Table 11.3.

Wheels provide an efficient mode of locomotion on smooth surfaces. Unfortunately, they perform poorly on rough terrain and offer moderate load-carrying capacity.

Legs, on the other hand, are highly maneuverable and convenient for negotiating rough terrain. On the negative side, they have limited stability and payload capacity. In addition, the plethora of actuators required to innovate joints leads to inefficiency.

Treaded tracks such as those used on military tanks offer high stability and payload capacity. However, they are slow, cumbersome, and inefficient. Perhaps

*Table 11.3* Modes of locomotion

| Mode | Advantage | Drawback |
|------|-----------|----------|
| Wheels | Fast on smooth surface | Poor on rough terrain, e.g., stairs |
| | Efficient | Moderate payload capacity |
| Legs | Good on rough terrain | High power consumption |
| | Highly maneuverable | Low stability |
| | | Low payload capacity |
| Tracks | High payload capacity | Low speed |
| | High stability | Low efficiency |
| | Fair on rough terrain | Detrimental to surface, e.g., grass, carpets. |
| | | Poor maneuverability |
| Rails | High payload capacity | Fixed route |
| | High stability | Expense of laying rails |
| | High speed | |
| | High efficiency | |
| | Simple to control | |

their greatest disadvantage lies in their destructive tendencies, wreaking havoc on delicate surfaces such as grass and carpets.

On the other hand, autonomous vehicles on guided rails offer stability, speed, efficiency, and high payload capacity. But they are obviously constrained by the fixed routes defined by the rails as well as the capital expenditure of laying the rails.

## Appendages

The structure of a robot must facilitate its interactions with the environment. In this section we explore the different configurations that may be deployed in robotic appendages.

The versatility of a limb depends on the number of ways it can move. Each *degree of freedom* refers to an independent axis about which physical movement can occur. Three degrees of freedom are required to locate an object in space: the chandeliers in a room might be located four meters from the western balcony, five meters from the southern windows, and three meters from the floor.

In general, physical objects need to be oriented properly as well as located in space. One does not generally care whether a golf ball rests on one of its dimples or another; but one could be partial about a wine glass being sideways, upside down, or right side up, especially if it contained something worthwhile. An object can be rotated about any of the three axes used to locate it in space. For this reason, six coordinates specify the position of an object: three for translation plus three for orientation.

Table 11.4 identifies the components of manipulators in terms of hands, joints, and limbs. The hand of a robot, also called a *gripper* or an *end-effector,* need not be anthropomorphic. Fingers are convenient for picking up eggs, magnets for handling metallic workpieces, and vacuum cups for lifting sheets of paper. A robotic hand need not grasp a welding torch; rather, the torch might be designed to fit directly into the "wrist" and serve in lieu of a hand. While we have the luxury of changing our wardrobes and accessories, a robot can enjoy the convenience of exchanging parts as well.

*Table 11.4* Components of manipulators

| Component | Design Considerations |
|---|---|
| Hands | Type of application, e.g., cooking vs. welding |
| | Touch vs. force transmission, e.g., tactile recognition vs. tightening nuts |
| Joints | Degrees of freedom required, e.g., materials transfer vs. writing script |
| | Work envelope required by the application |
| Limbs | Strength required |
| | Length of each limb |
| | Degrees of freedom required |

The design of a robotic *limb* depends on its envisioned uses, including the dexterity and strength required to accomplish the tasks. The structure of the limbs in turn defines the *joints*. More specifically, the articulation of each joint depends on the required functionality and the size of the workspace in conjunction with the configuration of the limbs.

## Articulation and Reference Frames

Our society is laid out in rectilinear fashion. Our streets intersect largely at right angles, as do our walls and the decor. Our thoughts are expressed in a linear sequence of words, and even our sense of time is linear as opposed to, say, circular. In this milieu, an obvious choice of a coordinate system is rectilinear, consisting of the three translational dimensions of depth, breadth, and height.

On the other hand, our limbs have a rotational rather than translational basis. A punch to the jaw might fly in a straight line, but the joints that drive it rotate rather than slide. The appropriate articulation of joints, and the reference frames that model the modes of movement, depend on the application.

Table 11.5 depicts a number of common configurations. In the *rectangular* configuration, all joints have a prismatic basis to permit the limbs to slide past one another. An advantage of this coordinate system is that the hand can traverse long distances accurately. The *work envelope* is given by the boundary of all the points in space which can be reached by the hand or end-point of a manipulator employing the corresponding configuration. The work envelope is obviously one factor in the selection of an articulation for a robotic limb.

The *cylindrical* configuration involves two translational degrees of movement and one rotational. Like the rectangular approach, this coordinate system leads to a simple work envelope.

The *spherical* arrangement consists of one prismatic and two rotational axes. It is a straightforward design that offers good lifting capacity.

Finally, the *revolute* configuration has all three rotating joints. It is a versatile design with a large work area.

The limbs of a robot provide for locating an object in space. As discussed previously, however, an object must also be oriented properly. This can be achieved in a straightforward way by providing for three axes in the wrist. The three degrees of freedom are known as roll, yaw, and pitch, as shown in Figure 11.3.

In addition, a mechanical hand should possess at least one more degree of control for grasping and releasing objects. This capability may take various forms, whether of motion in fingers, a magnetic switch, vacuum suction, or other implementations.

## Power

A manipulator or transportation module for a robot may be driven through various modes of power transmission. Table 11.6 lists three common types—electromechanical, hydraulic, and pneumatic—as well as their respective advantages and limitations. The first entry in the table indicates, for instance, that mechanical

*Table 11.5* Articulation of robotic limbs

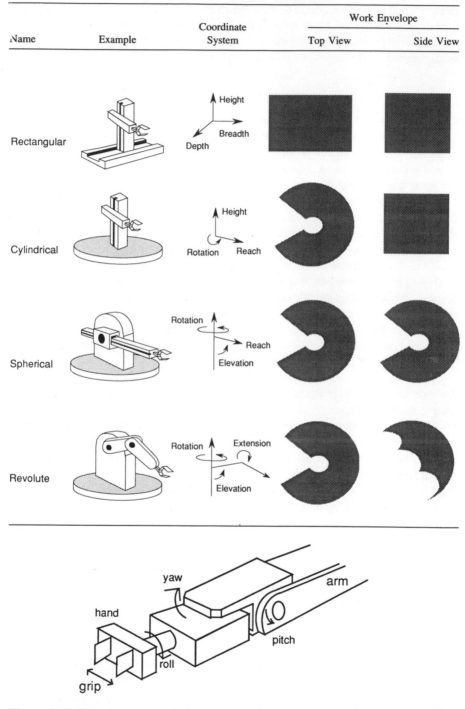

| Name | Example | Coordinate System | Work Envelope Top View | Work Envelope Side View |
|---|---|---|---|---|
| Rectangular | | Height / Breadth / Depth | | |
| Cylindrical | | Height / Rotation / Reach | | |
| Spherical | | Rotation / Reach / Elevation | | |
| Revolute | | Rotation / Extension / Elevation | | |

**Figure 11.3** Three degrees of rotational freedom in a robotic wrist, plus one for gripping.

*Table 11.6* Modes of power transmission

| Mode | Advantage | Drawback |
|------|-----------|----------|
| Electromechanical | Accurate<br>Convenient to service | Heavy |
| Hydraulic | Light weight<br>High power | Possibility of messy leaks<br>Need for external pump |
| Pneumatic | Light weight | Least accurate<br>Need for external pump |

linkages driven by electric motors are accurate and convenient. On the other hand, the gears and motors contribute significantly to the weight of a robot, and may strain the power source.

## Communication

A robot may communicate with humans or other machines through voice or radiation. Table 11.7 highlights the strengths and limitations of each mode. Even humans will soon come to wear portable radios or wristwatch videos for convenience in accessing and disseminating information, if some measure of privacy can be assured. Over the next few decades, at least, robots will have no qualms about privacy, and will likely incorporate all three types of communication listed in the table.

# Time

Temporal issues in robotic design can be broadly grouped into three classes: microlevel, unit-level, and macrolevel. *Microlevel* issues refer to realtime decisions concerning robotic activity. An example of a microlevel design issue is whether to incorporate sufficient physical power and reasoning capacity to enable a robot to

*Table 11.7* General-purpose modes of communication between the robot and human. Voice and radio are received as acoustic signals, while radio and television are transmitted as electromagnetic signals.

| Mode | Advantage | Drawback |
|------|-----------|----------|
| Voice | Convenient to humans | Limited range |
| Radio | Long range | Sometimes inconvenient to humans |
| Television | High information density<br>Long range | Sometimes inconvenient to humans |

avoid a falling obstacle or jump clear of a human rushing around a corner. Other examples of realtime issues were discussed in Chapter 6.

*Unit-level* issues refer to parameters that operate over the design life of the robot. An example in this category is the use of structural reinforcements in the body of the robot: the degree of physical enhancement will depend not only on the cost of production and operational expenses, but also on the severity of the application environment. A robot in a chemical plant may well suffer corrosive damage over time, but a search-and-rescue robot is subject to other hazards.

Temporal issues at the unit-level also include stochastic deviations in performance. The reliability demanded of a remote robot will be higher than that of a household aide; the availability or "uptime" for an industrial robot might be higher than an office machine; the servicing of a military scout should be speedier than that of a mail clerk.

The third class of temporal issues belong to the *macrolevel,* involving time scales connected to environmental trends. Included in this category are the nature and pace of advancing technologies as well as social values.

Table 11.2 has presented a reasonable timetable for the development of a number of robotic technologies. The table indicates, for example, that locomotion on two legs has already been demonstrated on research protoypes, but will not mature until the 1990's, leading to widespread commercialization after the year 2000. In a similar way, Table 11.8 forecasts the applications of robotic systems. For example, current efforts in artificial intelligence and mobile robots will lead to prototypes of a household companion that can perform chores and serve as a conversational partner. Full commericalization, however, is not likely to occur until the new millennium.

The social environment is difficult to monitor properly, let alone predict. With the natural human concern about losing control, some roles might become explicitly forbidden to robots. Today the computers that regulate electric power generation and dissemination cannot, for practical purposes, be turned off: to do so would risk not only economic and psychological hardships, but would also invite rioting in the streets.

In a contrasting application, the strategic defense network of the United States is highly automated and the relevant information is processed to a great degree before reaching the executive level. The President of the United States, as commander-in-chief, relies on this information. But if only a tiny fraction of the original information filters into the presidential suite, where is the real decision made concerning local conflicts or even the launching of nuclear missiles?

Despite the degree of automation, a robot, even if it displayed greater intellectual capacity and stability than any human, is not likely to be elected President in the forseeable future. But times can change: crises can temper the spirit and new generations of humans may grow to feel more comfortable with the products of their creation.

## Process

In this section we explore a number of generic issues in sensing and reasoning, including the coordination of numerous processes occurring concurrently in a robot.

*Table 11.8* Applications of robotic systems

| DOMAIN | Pre-1990 | 1990's | Post-2000 |
|---|---|---|---|
| **Industry** | | | |
| Production (stamping, welding, etc.) | ▬▬▬▬ | | |
| Materials handling | ▬▬▬▬ | | |
| Assembly | ⅠⅠⅠⅠⅠ▬▬ | | |
| Inspection | ⅠⅠⅠⅠⅠ▬▬ | | |
| **Office** | | | |
| Mail Handler | ▬▬▬▬ | | |
| Clerk | ⅠⅠⅠⅠⅠ▬▬ | | |
| Cleaning | ⅠⅠⅠⅠⅠ▬▬ | | |
| Professional | ░░░░ⅠⅠⅠⅠⅠ | | ▬▬▬▬ |
| **Home** | | | |
| Tutor | ⅠⅠⅠⅠⅠ▬▬ | | |
| Housekeeper | ░░░░ⅠⅠⅠⅠⅠ | | ▬▬▬ |
| Companion | ░░░░ⅠⅠⅠⅠⅠ | | ▬▬ |
| **Military** | | | |
| Automatic Pilot | ▬▬▬▬ | | |
| Scout | ░░░░ⅠⅠⅠⅠⅠ | | ▬▬▬ |
| Soldier | ░░░░ⅠⅠⅠⅠⅠ | | ▬▬ |
| **Ocean** | | | |
| Explorer | ▬▬▬▬ | | |
| Constructor | ░░░░ⅠⅠⅠⅠⅠ | | ▬▬▬ |
| **Space** | | | |
| Stationary Observer (on Mars) | ▬▬▬▬ | | |
| Rover (on Mars) | ░░░░ⅠⅠⅠⅠⅠ | | ▬▬▬ |
| Laborer (space station & moon) | ░░░░ⅠⅠⅠⅠⅠ | | ▬▬▬ |

░░░░ = Laboratory prototypes
ⅠⅠⅠⅠⅠ = First commercial applications
▬▬▬▬ = Widespread commercial applications

As usual, the topics are compartmentalized for convenience in discussion. However, we should keep in mind that many of the issues are interrelated, both within the realm of processes and in the context of other dimensions.

## Sensing

Figure 11.4 shows a taxonomy of sensors. The distal modes of sensing, whether in the form of sound, vision, or smell, can occur at a distance. The most common

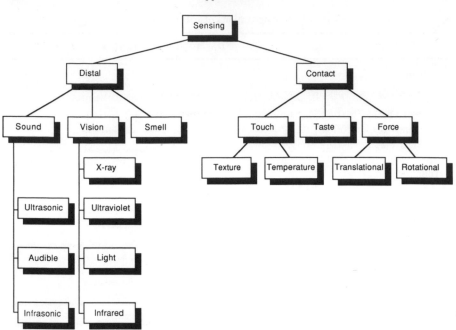

**Figure 11.4** Typology of robotic sensors.

sonic detectors may be those at frequencies audible to the human ear, namely, 20 to 20,000 cycles per second. Infrasonic frequencies may be helpful for detecting long-range tremors such as earthquakes or even passing vehicles. On the other hand, ultrasonic frequencies may find use in detecting irregular emissions in machinery such as motors, or for navigation in darkness when coupled with a corresponding mechanism for emitting the waves.

In a similar way, the most popular type of vision system is likely to be tuned to visible light. Infra-red, however, has its uses for detecting warm objects in the dark, or in identifying hot spots in machinery. Ultraviolet and X-ray vision may find more specialized applications in areas such as monitoring nuclear power generators. In collaboration with an external source of emission, an X-ray detector could find some use in search-and-rescue operations in areas dense with obscuring foliage or debris.

Olfactory sensors are in routine use as smoke detectors. They are also used to monitor ambient air in factories, such as microelectronic plants where poisonous chemicals often abound. Future uses will surely include the detection of unseen enemy materiel on the battlefield, and the regulation of fragrance in the household.

Sensors that require direct contact with the referent source are touch, taste, and force. Touch receptors may be helpful for identifying the texture of an object or for assessing the condition of old machinery. Touch sensors of two types are embedded in the human skin. One type is the velocity detector, which responds to the rate of skin indentation. This information is useful for determining the texture and shape of objects in contact. The second category is that of acceleration detectors which react to rapid vibrations. These sensors monitor, for example, the slippage of an object held in the hand.

Touch receptors in robotic hands can be implemented through piezoelectric sensors. When these devices deform under physical stress, they produce a voltage response in direct proportion to the applied pressure. In addition to physical motion, touch detectors may be used to gauge the temperature of an object. This may be done by monitoring the direction and rate of heat flow in conjunction with the nature of the material in contact.

When viewed as the analysis of compounds through chemical treatment, taste can be regarded as a routine capability. This sensing mode was invoked when the Viking landers on Mars retrieved samples from the alien soil and studied them through chemical tests. It also occurs in industrial plants in regulating the composition of processed fluids as well as the amount of pollutants in the effluent. Taste will surely be a requirement in the sensory modalities of a robotic chef.

A capability often given short shrift in cataloguing human sensory modes is that of force. This mode is often ignored or lumped implicitly under the category of touch. In the human body, however, the force transducers operate independently of the touch receptors. We use these sensors to monitor the translational force in lifting a weight, pushing a table, or inserting a thumbtack. They are also used to detect the rotational force or torque involved in wringing a towel, closing a jar, or turning a doorknob. Another important sensory mode relates to gravitational force, especially its orientation and its consequences for maintaining balance.

A robot will require most or all of these sensory capabilities too, if it is to share our space. The advantages and limitations of the respective sensory modalities are outlined in Table 11.9.

## Internal Monitoring

In addition to monitoring the environment, a robot must keep track of its own internal state. It must replenish its power pack before running out of energy, and replace a worn bearing before it damages an entire joint. The robot must also be mindful of poking itself in the eye or clobbering the left arm with its right. In fact, some of the early robotic prototypes had enough power that they succeeded in tearing themselves apart when given conflicting commands. Machines must be taught to abide by the Third Law of Robotics.

Monitoring the configuration of a limb may be achieved locally or globally. In the local mode, the angle of each rotary joint or the extension of each telescopic joint may be monitored and summed to yield a global value. The disadvantage of this approach lies in the cumulation of errors: an inaccurate reading in the attitude of a shoulder joint, when exacerbated by errors in monitoring the elbows, wrists, and fingers, may lead to unacceptably large discrepancies for delicate operations.

In the global approach, the location of an endpoint could be determined by triangulation with respect to the head and torso. The closed-loop configuration between the actual and desired positions of the fingers leads to enhanced accuracy. The drawback of the procedure is that it only works for objects in clear view; when a hand is behind the back or around an opaque object, triangulation is useless.

The human body uses both local and global approaches. We have some sense of

*Table 11.9* Modes of sensing

| Mode | Advantage | Drawback |
|------|-----------|----------|
| Vision | Standard form of human communications, e.g., signs, books | Direction-specific |
|  | Long range | Hypervision (e.g., infrared) required in absence of light |
|  | Infrared and ultraviolet frequencies can yield additional information |  |
| Sound | Standard form of face-to-face communications | Short range (for conversation) to medium (for alarms) |
|  | Omnidirectional | Poor directional resolution |
|  | Useful for exceptional conditions, e.g., sequealing motor. |  |
| Smell | Omnidirectional | Short to medium range |
|  | Useful for exceptional conditions, e.g., smoke or poisonous vapors | Limited number of uses |
| Taste | Useful for determining chemical composition | Requires contact |
|  |  | Limited number of uses |
| Touch | Useful for object recognition | Requires contact |
|  | Can determine temperature |  |
|  | Useful for vibrations |  |
| Force | Required for manipulating objects | Requires contact |

the configuration of each limb; often we can comb our hair or slip into our shoes without looking at them. For more delicate tasks such as threading a needle or writing a note, we tend to rely on a closed-loop configuration by watching as we perform each action. A sophisticated robot will also have multiple modes of sensing and control at its disposal.

## Coordination

A mobile robot is a rich assemblage of coordinative systems. These range from the high-level functions of the reasoner in managing major subsystems, all the way down to the control of individual joints.

Figure 11.5 shows a closed-loop control system for the force required to grasp an object. The desired level of force is fed into the control module, which compares it with the actual amount of force as indicated by the feedback signal. The discrepancy enters the command generator, which determines the direction and extent of adjustment necessary. The resulting command passes into an amplifier which produces power proportional to the level of the input signal. The power drives a motor attached to some linkage such as a set of gears. The mechanical linkage in the robotic hand ultimately converts the initial command signal into displacement at the fingertips.

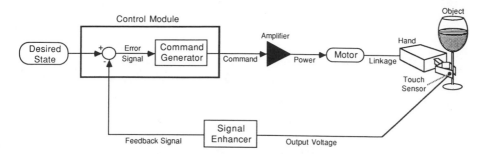

**Figure 11.5** Closed-loop control of grasping force for an electromechanical system.

Once contact is made with an object, even minute displacements in the fingers can correspond to significant levels of applied force, leading to slight deformations in the hand. These deformations are monitored by a touch sensor that converts spatial distortions into corresponding voltage levels. The output voltage, however, will be highly uneven due to noise from various sources such as slippage in the fingers or vibrations caused by passing vehicles. The task of the signal enhancer is to eliminate irrelevant spikes in the voltage and pass on a smooth signal to the control module. The entire robot will be rich in control systems both to manage its own subsystems as well as keep track of external parameters such as distances to objects.

## Ranging

Techniques for locating objects in space may be partitioned into active and passive modes. In the *active* mode, a signal is emitted from the ranging source and detected after being reflected by the target object; the elapsed time depends on the intervening distance. The ranging medium may be sonic or radiant. Sonic devices can readily measure distance over several meters to an accuracy of 1 millimeter. A disadvantage of using sound is the rapid attenuation over distance, thereby limiting the range of detection. A more effective source is coherent radiation such as a laser. The development of detection devices having a cycle time of a picosecond[12] will result in instruments with an accuracy of plus or minus 1 millimeter, even over distances of many kilometers.

A *passive* approach to ranging requires only the monitoring of a target, without the deliberate emission of signals for this purpose. The standard technique in this category is triangulation, in which two or more sensors focus on a target; the angles formed by the sensors with respect to each other can be used to determine the distance to the object (see Figure 11.6). Light and other radiation are most amenable to this technique. The resolution of sound waves, on the other hand, is too coarse to permit accurate ranging for the limited baseline between sensors on a robot, a distance short of a meter. As we have seen, sound waves can be useful for judging depth; but their poor transverse resolution renders them ineffective for gauging lateral displacement.

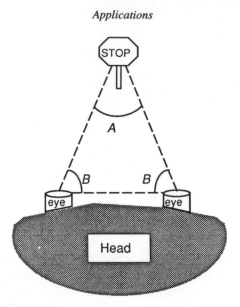

**Figure 11.6** Locating objects in space through triangulation. The distance to an object is determined by the apex angle *A* or equivalently by the base angle *B*.

## *Reasoning*

The purpose of the reasoning module is to discharge the responsibilities of the robot by integrating internal knowledge with external information. The reasoner must pursue its goal despite physical obstacles and informational deficiencies.

The robot may have to deal with unforseen circumstances as well as incomplete information. In addition, even the information that it has may be uncertain or contradictory.

The reasoning component must assess the state of the environment, including recognizing objects and detecting both rapid and gradual changes in the vicinity. The module must plan its strategy and implement its actions, whether in terms of locomotion, object manipulation, or communication with other agents.

In order to understand the environment, the reasoner must be able to recognize the *types* of objects. These objects may be physical, as in the case of people, vehicles, trees, and machines. Or they may be abstract, as in letters, symbols, fires, and potholes. The module must also understand the salient *properties* associated with particular objects, whether in terms of size, shape, color, texture, intensity, or other parameters. Although we as adults are accustomed to our daily environment, the world is a bewildering array of sensory images to an ingénue robot.

To make matters worse, objects do not reside independently in space. They interact in a myriad of ways, both directly and indirectly. Some examples of physical *relationships* between two objects are *above, below,* and *inside.* Furthermore, two objects or events may have temporal relationships: one causes the other, or results from it, or influences it in some other way. These effects may occur sporadically, always, or rarely. We humans seldom understand relationships well, even when they are largely of our own making, such as the interaction between two people or the impact of taxation on economic recovery.

The interpretation of sensory input and techniques for reasoning are topics of continuing research in artificial intelligence. Some progress has been made to the point of developing knowledge-based systems in various domains such as medicine, business, law, and engineering. The most difficult capability to computerize is what we call *common sense:* without being told explicitly, we avoid fires, refrain from stepping out of windows, wear a shirt before putting on a coat, and seek rest when fatigued.

Contrary to the expectations of researchers at the infancy of artificial intelligence in the late 1950's, human expertise in specialized domains is much easier to convert into software than general knowledge. The skills of the physician can be automated before the talents of the tomboy. The key source of difficulty may lie in the fact that what we refer to as "common sense" is not a single skill but a rich spectrum of capabilities. Whatever may be the true nature of this beast, it too will surely be harnessed at some point and replicated in software.

## Efficiency

Efficiency considerations for a robot may be broadly classified into costs of production and operation. Sometimes these factors are largely independent, as in the case when two motors of equal performance and reliability differ only in acquisiton price. In an analogous way, if the torso can be made of two satisfactory materials which differ only in weight, then the lighter one will lead to reduced energy consumption in routine operation. Many times, however, the costs of fabrication and operation are at odds. Durable materials that lead to decreased maintenance costs will often cost more than those of weaker constitution. Redundancy in actuators may reduce service requirements but increase production expense. There is no universal formula for making such decisions, since they depend not only on the application but the precise values of the costs involved.

A perennial question in product design is the relationship between efficiency and performance. The production cost may be lowered by using a smaller motor, but the performance might be compromised.

The decision is influenced not only by the technical considerations but by economic factors and social expectations. An example of a financial factor relates to the prevailing interest rates. A consumer may be willing to pay $100 more today in order to save $10 over the coming year if the interest rate is 5 percent; but the decision is reversed if the rate is 15 percent.

An instance of technological expectations relates to the perceived rate of product obsolescence. Robotic equipment depends critically on computing technology, and will therefore be paced in part by advances in processing hardware and software. As a result, the *intelligence* in a robot, if not its structural hardiness, is likely to advance on a regular basis: a new generation of abilities will likely emerge every few years, coupled perhaps with decreasing cost. In this situation, a consumer may

**Figure 11.7** The Household Aide on an errand.

balk at paying twice as much for a robot that will last four times as long, say twenty years as opposed to five.

A factor related to product capability is quality. In the 1950's, a representative consumer was probably delighted to own an automobile, even if it broke down incessantly and spent half its time in a repair shop. Three decades later, the same consumer is not likely to be sanguine about a car that suffers a single major malfunction in the first years of ownership.

The robotic market will be subject to similar macroscopic trends. For example, if an extra design feature raises production cost but doubles product reliability, then the consumer may well be willing to pay a premium over and above the vendor's expense.

Robotic design, perhaps more than any other field, will be tempered by social and psychological as well as financial factors. Given the similitude between autonomous machines and people, human factors in all aspects will figure prominently in the construction and deployment of robots. In this milieu, efficiency will be but one parameter of design.

## Summary

Intelligent machines can provide many advantages over traditional equipment or human labor. Robots can work tirelessly with rapidity and efficiency. They can deliver enormous power or provide a gentle touch as needed. They can be as accurate in locating objects in space as consistent in behavior.

To date, most robotic applications have been in tasks that are unsavory to humans. These include the monotonous jobs on the factory floor or dangerous tasks such as defusing bombs. Robots have gone where humans cannot yet venture: to the depths of the oceans and the outer reaches of the solar system.

In the years ahead, autonomous robots promise to be as helpful in the home as in the office, seabed, or factory floor. Figure 11.7 is a preview of things to come.

# 12

## Flexible Factory

*Among civilized and thriving nations, on the contrary, though a great
number of people do not labour at all . . . . the produce of the whole
labour of the society is so great, that all are often abundantly supplied.*[1]

Adam Smith

*There is a minimum number of parts below which complication is
degenerative, in the sense that if one automaton makes another the
second is less complex than the first, but above which it is possible to
construct an automaton of equal or higher complexity.*[2]

John von Neumann

The increasing rate of technological innovation, product obsolescence, and market
segmentation in the world economy calls for a new generation of production facilities. The implications for a modern factory are challenging. In contrast to the huge,
special-purpose factories of yesteryear, the future factory must combine the flexibility to manufacture a wide variety of products in small lot sizes with high quality
but low cost. Further, new product lines should be introduced at a moment's notice,
and old ones discontinued at will.

A flexible factory is an adaptive system that can manufacture several products
concurrently in a wide range of lot sizes. The ability to accommodate a spectrum of
products is attained by configuring the hardware resources of the factory into a set
of virtual production lines. Each virtual line is an abstract sequence of operations
that may be implemented on one or more physical devices.

In this approach, the factory is reminiscent of a mainframe computer system that
masks the limitations in primary memory by accessing secondary storage in a
process invisible to both the user and her programs; moreover, the physical machine
in conjunction with its supervisory control program can accommodate several users
in parallel, each interacting with a virtual computer configuration tailored to individual needs. Through the ethereal architecture superimposed on the hardware
resources, the flexible factory can also adapt rapidly to dynamic market conditions
by accommodating fluctuating lot sizes and new product introductions.

This chapter explores the design implications for the next generation of manufacturing automation technology which should be feasible by the turn of the

century. The domain of production is for discrete parts manufacturing, such as bottles or cars, rather than continuous production such as beer or gasoline.

The design considerations and system architecture should be portable across a spectrum of industries. The main distinction between an implementation in one industry or company versus another will be the tailoring of domain-dependent modules and their attendant knowledge bases.

## Purpose

An intelligent production plant should be able to manufacture a variety of products in differing batch sizes. In this chapter we explore the design issues for a flexible factory whose objectives are defined by the following functional requirements:

1. Manufacture from 1 to 1,000 products simultaneously.
2. Accommodate lot sizes from 1 to 1,000,000.

A number of constraints also apply to the design of the factory. These constraints define the boundaries of acceptable solutions.

1. *Responsiveness*. The system should reconfigure for a new product within 1 second.
2. *Inventory*. Average inventory levels should not exceed one day's usage.
3. *Learning*. The plant should monitor the history of its operations and discern ways to redesign a product or its production process to drive down production costs over time.
4. *Self-repair*. The factory should accommodate routine breakdowns and engage in self-repair.
5. *Efficiency*. The plant should be competitive in initial construction and routine operation. For example, only a handful of human technicians should be on call to handle non-routine breakdowns.

The first constraint specifies that the factory should prepare for a new product at a moment's notice. This reconfiguration is a conceptual or virtual activity rather than a physical task involving the relocation of equipment. These ideas will be clarified in a subsequent section dealing with the notion of a virtual factory.

The main advantage of a flexible factory is its versatility. In performing physical work in conjunction with information processing, it can be programmed for a variety of tasks. Precise machine operations ensure product quality, while automated manufacturing cells enhance efficiency through rapid tool changes, workpiece setups, and processing operations. Further, the integrated supervisory system coordinates information flow and provides for managerial control.

## Space

The environmental considerations for factory automation are manifold. These range from technological to social and economic factors. The likelihood of successful automation efforts depends as much on healthy economic conditions as on managerial commitment.

## Environment

Flexible automation is driven by a number of socio-economic factors. Developers and automation technology producers want funding and facilities; users are interested in a business climate (for example, tax and trade laws) that is favorable to the introduction of flexible automation; members of the labor force seek employment stability, better jobs, and improved relations with management; state and local governments attempt to promote both economic expansion and a healthy employment base. Educators and trainers require funding, equipment, and facilities, as well as curricula that are responsive to changing technology. Additional demands include those of the federal government, which needs flexible automation for the defense industry and wishes to promote the ancillary effects of the technology on productivity, economic growth, and employment.

The imperative of capital investment in production facilities springs from international as well as domestic influences. The worldwide commitment to automation research, such as efforts in Japan and Germany, makes it all the more important for the United States and other countries to follow suit with technologies for cost reduction.

As a result of the lead taken by the defense industry in flexible automation and the introduction of intelligent systems, overall manufacturing capacity has increased in North America. Major contributors to research in the United States are such governmental institutions as the Department of Defense, National Aeronautics and Space Administration, the National Science Foundation, and the National Institute for Standards and Technology. These organizations, by funding project development, act as catalysts for the introduction of novel technologies and methodologies into the country's private sector.

## Technology

The flexible factory will depend on a variety of technologies. Computer-aided design (CAD) allows for computer representation of the product and for the preliminary analysis of product performance and manufacturability. Machinery such as robots and numerically-controlled (NC) machine tools perform the physical operations, while knowledge engineering or artificial intelligence techniques allow for comprehensive monitoring, fault management, and error recovery. Clusters of machine tools under computer control, known as flexible manufacturing cells (FMC's), allow for the automation of manufacturing processes which are coordinated through the hardware and software within a larger computer-integrated manufacturing (CIM) environment.

The flexible factory will depend on a spectrum of technologies for design, scheduling, fabrication, and software development. A historical profile of these methodologies is outlined in Figure 12.1. A set of projections for some critical technologies under development is given in Table 12.1.

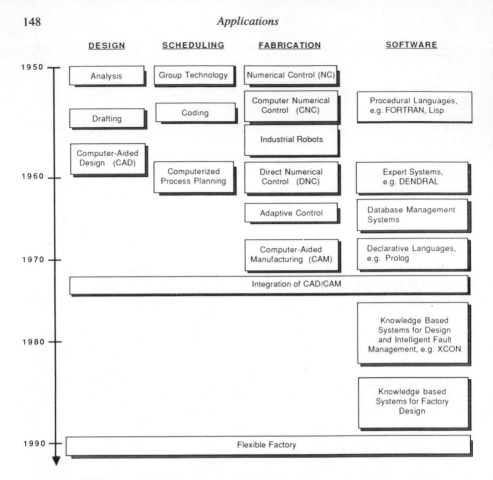

**Figure 12.1** History of automation technologies for the manufacturing environment.


A computer-integrated production system is especially effective for low-to-medium scales in the diversity of products as well as production rate. Low product variety and high production volumes indicate the need for dedicated production lines, while high levels of variety and low quantities suggest the use of stand-alone numerical control machinery. These regions of applicability are shown in Figure 12.2.


## Structure

The flexible factory consists of physical objects and the knowledge needed to direct them properly. The physical components of the plant are of the following types:

- *Operating equipment:* machine tools and related equipment to perform the manufacturing operations. These include general and specialized machinery as well as their tools.

*Table 12.1* Projections for some key automation technologies

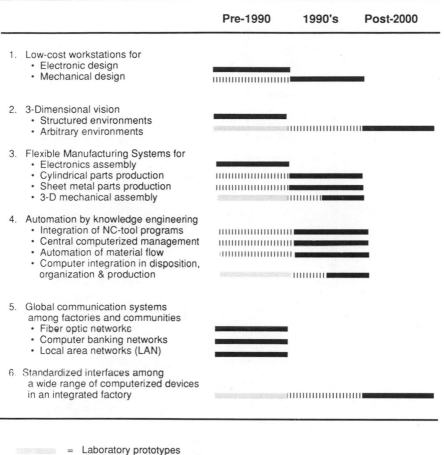

|  | Pre-1990 | 1990's | Post-2000 |
|---|---|---|---|

1. Low-cost workstations for
   - Electronic design
   - Mechanical design

2. 3-Dimensional vision
   - Structured environments
   - Arbitrary environments

3. Flexible Manufacturing Systems for
   - Electronics assembly
   - Cylindrical parts production
   - Sheet metal parts production
   - 3-D mechanical assembly

4. Automation by knowledge engineering
   - Integration of NC-tool programs
   - Central computerized management
   - Automation of material flow
   - Computer integration in disposition, organization & production

5. Global communication systems among factories and communities
   - Fiber optic networks
   - Computer banking networks
   - Local area networks (LAN)

6. Standardized interfaces among a wide range of computerized devices in an integrated factory

░░░ = Laboratory prototypes
||||||||||| = First commercial applications
▬▬▬ = Widespread commercial applications

- *Material handling system:* equipment to convey workpieces from one station to another.
- *Computers:* a local network of computer hardware to handle distributed processing activities as well as overall coordination of the virtual factories.
- *Human workers:* the final recourse for handling exceptional conditions. Human workers include computer programmers to fix software problems or implement major new capabilities. The interface to the environment presents perhaps the most complex and unpredictable problems; hence a loading supervisor will likely be required for receiving and shipping operations.

The factors to be considered in equipment selection are listed in Table 12.2. This exhibit is followed by the functions of the computer system in Table 12.3 and the role of human workers in Table 12.4.

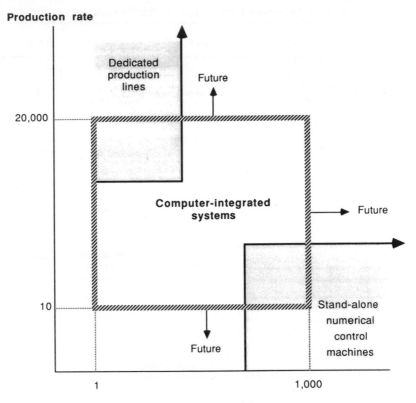

**Figure 12.2** Realms of relevant technologies in early 1990's, in terms of production rate and number of products manufactured concurrently.

*Table 12.2* Factors in equipment selection

---

Products to be manufactured
  Materials
  Size
  Shape

Lot sizes
  Large lots imply increased equipment capabilities

Product life span
  Long life suggests high-quality machine tools

Product variety
  Large variety suggests highly flexible equipment

---

*Table 12.3* Functions of the software modules

Design
  Graphics
  Design principles
  Simulation tools
  Others

Production
  Computer Numerical Control (CNC)
    Low-level control of machine tools
    More versatile form of Numerical Control (NC)

  Direct Numerical Control (DNC)
    Superset of CNC
    Storage of CNC programs, and distribution of pro-
    grams to individual machines

  Robotic operations

  Production management
    Process planning and scheduling
    Material requirements planning: ordering input ma-
    terial based on production requirements and inven-
    tories
    Fault management: monitoring and error recovery

  Traffic control
    Control of individual shuttles and conveyors
    Regulation of traffic among shuttles

Management information
  Machine utilization rates
  System performance: output by product type, prob-
    lems, etc.


*Table 12.4* Human workers in the flexible factory

System manager
  Has overall responsibility for factory operations
  Responds to unexpected deviations in production schedules
  May authorize changes in hardware or software
  Supervises other people
  Need not be on site 100 percent

Electrical technician
  Corrects electrical problems
  Must be on site 100 percent

Mechanical technician
  Handles hydraulic, pneumatic, and mechanical components
  Must be on site 100 percent

Loading technician
  Monitors potential problems at shipping and receiving
  Must be on site 100 percent

Computer Programmers
  Correct software faults
  Implement major changes in software
  Need not be on site 100 percent

The hardware and software components of the automated production environment may be depicted as a hierarchical structure as shown in Figure 12.3. Such a conceptual organization is, of course, independent of the physical realization of the system. In fact, an appropriate network architecture for the automated factory is given in Figure 12.4. This network configuration represents the communications structure for the differing components of the factory environment.

## Time

A key benefit of a flexible factory is its timely response to changing production requirements. These rapid responses relate both to minimal throughput times for existing product lines as well as rapid reconfiguration of the factory.

In addition to routine operation, the system incorporates capabilities for fault management. This is achieved by monitoring status information for trends that suggest impending problems. If a machine tool begins to vibrate at an uncharacteristic frequency, for example, then the monitoring module can request a self-diagnosis from the machine, suggest a change in the tool piece, reassign the current job, or take some other corrective action.

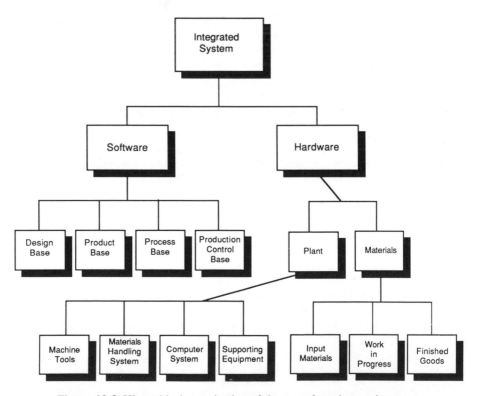

**Figure 12.3** Hierarchical organization of the manufacturing environment.

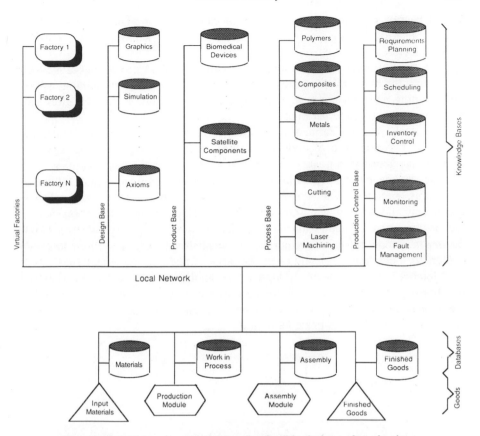

**Figure 12.4** Network configuration for flexible design and production.

When a problem does occur, the monitor identifies the problem, isolates the problematic components, and calls for the fault management system. The fault manager responds at once to prevent or curtail any damage to the workpiece or equipment, then invokes an error recovery module to redress the situation. The specific remedy will of course depend on the nature of the equipment, the fault, and the workpiece.

## Process

### Virtual Factory

The automated factory maintains its flexibility by the effective coordination of numerous processes. Superimposed on the physical factory is a set of abstract constructs called *virtual factories*. A virtual factory is defined by a sequence of production operations implemented on the machinery of the physical plant. Each virtual factory supports the manufacture of exactly one product or output of the physical factory. Its configuration is defined by the specifications of the product to be manufactured.

The configuration for a set of virtual factories is illustrated in Figure 12.5. In this figure, the first virtual factory is defined by four operations: machining, assembly, painting and packaging. In contrast, the second virtual factory requires only machining, painting and packaging.

A virtual factory is defined by a sequence of processes rather than machines. Therefore two consecutive products or units that are produced by some virtual factory may actually be treated on different machinery in the physical plant.

The advantage of the virtual factory is conceptual simplification in development as well as operations. The flexible factory is constantly subject to fluctuating requirements and novel product introductions. The logical separation of one virtual factory from another allows for the decomposition of complexity. When a virtual factory faces new production requirements, or malfunctions, or is eliminated altogether, the operational capabilities of the other virtual plants are unaffected.

Moreover, the separation of logical operations and physical devices further partitions complexity. Changes in the hardware configuration have little impact on the logic of production operations, whether a specific machine is modified, breaks down, or is joined by duplicate hardware.

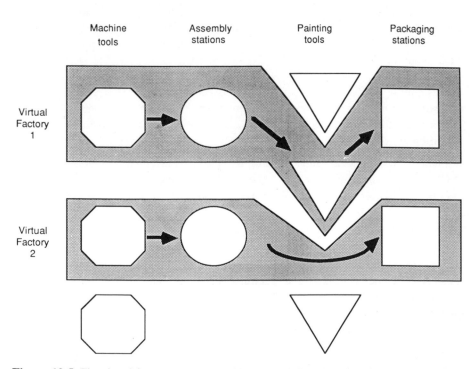

**Figure 12.5** The virtual factory: a conceptual structure defined by a set of production operations that may be effected on one or more physical devices.

## Centralization

The flexible factory combines elements of both centralization and decentralization. For example, the process plan for each product is determined by a specific virtual factory, but the scheduling is performed dynamically by individual machine tools based on the priorities of differing jobs. These priorities may be determined in part by hard constraints such as a human agent who stipulates, "Ship 5000 laser disks by the end of the week." The priorities are also influenced through soft constraints by the virtual factories which may bid for equipment time for specific operations.

In a conventional computer system, the virtual devices and operating systems are orchestrated by a master operating system. For example, the top-level operating system determines which program has access to a processor or a disk drive at a particular time. The complexity of an automated factory, however, would render such a centralized scheme infeasible. Rather, the control would have to be distributed among the virtual factories. One potential mechanism for distributed control is a market economy where each virtual factory bids against others for access to a particular machine tool.[3] In this scheme, each factory is assigned an "income" of credits which it may use to purchase time on specific equipment, or otherwise negotiate with other agents.

## Determinism

The flexible factory is non-deterministic in the sense that a given job or order is not assigned a specific set of machines *a priori*. The particular machine assigned to a painting operation, for example, will be a dynamic decision based on the utilization of the painting cells as well as the needs of other jobs.

The reliability of the factory is enhanced by the separation of the logical processes and the hardware devices. Their relative independence implies that a failure in a software module will have little, if any, effect on the hardware, and vice versa.

Further, the operation of concurrent virtual factories implies that a failure or modification in one production line will occur independently of the others. System reliability is further enhanced by the intelligent diagnostic system which attempts to give advance warning of problem areas, and the error recovery module which compensates immediately for mishaps when they do occur.

# Efficiency

An important goal of the intelligent factory is to reduce costs by increasing efficiency. The efficiency results from a spectrum of sources, ranging from the substitution of machine time for human labor, the careful use of material inputs, high-speed operations, increased capital utilization, and decreased inventory levels.

## Development Costs

It is unlikely that a facility such as the flexible factory will be available for sale as a turnkey system by the end of this century. Although many of the hardware components may be available as machine tools and flexible manufacturing cells, the large-scale integration issues will still have little precedent by then. As a result, such factories will have to be tailor-built to the needs of specific organizations.

For such a novel enterprise as the flexible factory, system development concerns may account for perhaps half the cost of the facility, the remainder being attributable to off-the-shelf hardware. Although the system will be directed by software running on hardware, the computing devices will comprise only a fraction of the overall cost of intelligence. In the 1990's, the cost of developing a typical knowledge-based system will likely comprise 10 percent for computer hardware and 90 percent for software. More specifically, this 90 percent figure is divided between prototype development (20 percent), and maintenance (70 percent) to accommodate changing system requirements over time. The disproportionate cost of software will persist until the artificial intelligence community develops better techniques for automatic programming and autonomous learning.

The integrating software for the automated factory will require hefty amounts of new programming, perhaps over tens of millions of lines of code. A look at software engineering over the past two decades indicates a remarkably uniform rate of productivity for engineers working on large projects: about 10 to 20 lines of code per day per programmer, including the generation of associated documentation, at a cost of roughly $15 to $20 per line.[4]

These figures are relatively independent of the application domain and the type of language used, whether a missile control program written in assembly language, or a management information system in COBOL. If these rates of programming productivity hold for plant design, then an operating system containing 100 million lines of code will require roughly 40,000 programmer-years and over $1 billion.[5]

In this estimate, the cost of software development is about an order of magnitude higher than the hardware cost of building an average factory without intelligence. For the sake of comparison, a representative price for a traditional factory (a plant to manufacture computer chips, for example) is a good fraction of $1 billion; of this cost 70 percent is dedicated to equipment.[6]

Unlike hardware, however, software can be readily duplicated. Therefore the development cost of an operating system can be amortized over any number of virtual factories, and therefore entire physical plants.

Given the difficulty of managing large software projects, how feasible is it to manage 40,000 programmer-years? If the project could be partitioned into 10 departments or subcontractors, each with 50 teams, and the interfaces among program modules clearly defined *a priori,* then 500 teams of 10 programmers each could complete the task in eight years. As a practical matter, it is unlikely that the interface could be clearly specified in full beforehand. One obvious strategy would be to build a series of small prototypes before embarking on the main project.

Another strategy to enhance productivity is to use higher-level languages wherever appropriate. For example, the representation of decision rules in Prolog may shrink programming effort by an order of magnitude in comparison to the use of Fortran, which in turn is an order of magnitude more powerful than assembly language. Once the factory operating system is in routine use, then the most heavily-loaded modules may be rewritten in a more efficient medium such as C or perhaps even assembly language.

## Conclusion

In the past, attempts to implement factory automation have stumbled for various reasons. Among the major hurdles are a dearth of experience and lack of standards, both in terms of software development and interfaces between devices.

Another stumbling block is inertia. A comprehensive, fully-integrated system is the most efficient form for automation, since an integrated plant is faster and reduces work-in-process inventory investment. However, a completely integrated system is often seen as too expensive or too risky to install.

Yet another factor relates to personnel issues. As with any major technology, factory automation has both positive and negative implications for workers. Until the eighties, U.S. labor and management were often at odds over the introduction of new technology. Despite a long history of distrust between the two groups, the decline of American economic competitiveness in world markets has highlighted the need for cooperation in implementing productive equipment which will ultimately result in a higher standard of living for everyone.

In the past, workers have often been wary of being controlled or even displaced by new technologies. However, increased productivity and industrial competitiveness due to automation may actually prove to increase overall demand for human labor. Although changes occur in the types of tasks people perform, the demands for engineers, computer scientists, technicians, mechanics, repair and installation experts tend to increase. While the need for craftworkers, operatives, and laborers may fall, demand for upper-level managers, technical sales, and service personnel will rise. Basic literacy, important for all jobs relating to automation, will be necessary as jobs come to demand analytical and problem-solving skills. Due to intelligent machines incorporating self-diagnosis and other abilities, the variety of required skills will increase, while the amount of judgment and the time needed to become proficient in the execution of these jobs should decline.

Management will experience several changes as a result of the introduction of new technology and its impact on workers. Changes in the work environment could result in boredom due to the lack of challenge in the operation of flexible machine cells and other automated equipment. Computer automation can cause tension; the pace is set externally and the complexity of the system requires a high degree of sophistication for equipment repair—both these functions will be out of the worker's control and may result in discontent. Pressure to avoid the high opportunity cost of downtime in equipment, the potential for computerized supervision, and the fatigue and possible danger of long exposure to video display screens will require active efforts to avoid a potentially harmful environment. All of these changes underscore the need for improved dialogue between managers and workers.

Various policies may be adapted to promote efforts toward flexible automation while remaining mindful of the human element. A combination of technological growth and human development will be required to promote flexible automation and industrial productivity while minimizing the negative effects of technological advances. This may be achieved through efforts in education and training.

There is a pressing need to redress the historical neglect of manufacturing processes, organization, and management. Potential remedies include the promotion of standard-setting efforts, provision of technological information to encourage its use, updating of university curricula, and the formation of a national agency or public corporation to address these goals.

For workers, there is the need to implement a long-term program for the shift in labor demand resulting from new technology as well as regional dislocation as the foci of economic growth shift from one area of the country to another. Further efforts are needed to establish job creation programs; possible solutions include the creation of jobs through increases in public goods and services as well as the reduction of the workweek coupled with an increase in hourly wages. Another mechanism is the provision of tax incentives to relocate personnel within existing firms. An expansion of labor-market information would provide a means to measure change and aid both labor and management in adjusting to production reorganization.

A federal work environment policy is needed in order to analyze the social and physical effects of automation in the workplace. Congress can promote the standards to be set for automated workplaces by federal health and safety agencies.

A broad approach will be necessary for all workers and managers to understand their situation and to enable them to better respond to change. To this end, a federal policy toward education, training, and retraining must be established to renew the emphasis on basic skills necessary for work with automation technology. Tax incentives and grants for curricular development and instruction—including provision of standards for automation education, in-house instructional programs, and self-education—would facilitate a smooth transition.

## Summary

This chapter has explored the nature of an advanced factory that should be practical by the end of this century for progressive organizations. The flexible factory accommodates rapid changes in production requirements or even new designs through the concept of the virtual factory. A virtual factory is an abstract structure defined by a set of manufacturing operations relating to a specific product. This logical structure, superimposed upon the production equipment of the physical factory, can operate in parallel with hundreds of other virtual factories.

The logical separation of equipment and processes allows for flexibility in production scheduling, which is performed dynamically. The system is also robust, since the errors in hardware are logically isolated from those in software. Further, one virtual factory cannot interfere directly with another. Their only mode of interaction is indirect, by way of the distributed operating system whose job is to assign equipment resources to specific processes.

Many of the underlying technologies for the flexible factory, such as autonomous transport systems and flexible manufacturing cells, are already at hand. Other capabilities, such as learning systems, are in their infancy. But progress even in these areas continues steadily, resulting from concerted research efforts in areas such as neural networks, which learn to recognize new patterns with little or no supervision.

Perhaps the most challenging barrier will lie in system design issues. As a society we have acquired experience in constructing intelligent devices—such as electronic chess players—and even more experience in building complex structures such as spacecraft and legal codes. However, the task of building large-scale systems incorporating sophisticated behavior is new to us.

Despite the paucity of experience, we are beginning to acquire an intuitive idea of effective ways to combine a multitude of components that interact toward purposive ends in meaningful fashion. The general framework for intelligent systems is intended to serve as a reference point. This framework highlights the critical issues to be addressed, from the uppermost reaches to the lowest levels of a complex system. Through reference to the framework, the design effort may proceed in a more coherent, systematic fashion.

# 13

## Organizational Design

*Disorder leads to what Prigogine terms a bifurcation point at which
the organization can either crumble into chaos, as many businesses
have, or jump to a new, higher level of order. . . . Apple [Computer]
had reached the bifurcation stage during its severe downturn. Our
chaos, however, led to the creation of a new, different, and stronger
entity whose roots still remain intact.*[1]

John Sculley

The Apple Computer Company is a widely-recognized corporate success story. It
grew out of a partnership in 1975 between Steven Jobs, the visionary iconoclast,
and Stephen Wozniak, the technical wizard. The company had begun with a new
personal computer designed by Wozniak for the hobbyist and household user.

Jobs had recognized the business potential of the new computer and arranged for
its production out of his parents' garage in 1976. Jobs provided the entrepreneurial
energy: he cheered on Wozniak, masterminded the packaging of the product, and
developed the marketing plans; but perhaps as importantly, he succeeded in attract-
ing Armas (Mike) Markkula, who contributed managerial experience and business
acumen.

The three founders were a study in contrasts. Markkula, the businessman, hoped
to augment his multimillion-dollar fortune through the new venture. But he
cherished his privacy and sought to remain backstage, away from the limelight.
Wozniak, the engineer, sought neither fame nor fortune. He was, and wished to
remain, a tinkerer; his second ambition was to become a school teacher. In the
meantime, however, he was happy in a secure job at Hewlett-Packard, and wary of
joining a start-up company for fear of becoming embroiled in managerial politics.
Only concerted pressure from Markkula and Jobs finally moved Wozniak to partici-
pate on a full-time basis in the new company.

Jobs, on the other hand, sought personal growth and business experience. He
emerged as the company spokesman, a visionary who sought to put an Apple in
every home and office at a time when no other expert or layperson could envision a
personal computer business, let alone a whole industry.

In the second half of the seventies, the United States was still shuddering from
the shock of the oil crisis of 1974. The world economy was beginning to recover
from the imbroglio, but American business still lacked confidence due to stiff

competition from foreign firms in both domestic and international markets. The country badly needed heroes to restore its confidence in the free enterprise system and in the ability of U.S. firms to regain a competitive edge.

Who better to fill the role than Steve Jobs, the energetic, magnetic personality from Silicon Valley, the embodiment of the garage-to-riches dream, the epitome of American ingenuity and free enterprise. Americans needed heroes, Apple needed free publicity, and Jobs needed a role. It was an ideal match, one which later was to spark a crisis of epic proportions at the company. For the man who had become the symbol of Apple Computer would not readily change his position, even when differences of opinion within management ranks led to employee tension and depressed performance.

Apple Computer, Inc., was formed on January 3, 1977. Based on the Apple II computer, the company grew at a breathtaking pace and earned the distinction of being one of the fastest-growing firms to reach billion-dollar status.

After an extensive search, John Sculley was hired in April 1983 as the third president and chief executive officer of Apple Computer. He took on the post held briefly by Mike Markkula after the departure of the first president, Michael Scott. Despite a childhood interest in technical matters, Sculley had pursued more conventional pathways. He had spent a decade and a half in the soft drink industry, during which he rose to the position of president of Pepsi Cola USA. He had earned acclaim as the marketing star who masterminded the "Pepsi Generation" theme, an advertising campaign that led to Pepsi's co-leadership in market share with its archrival Coca-Cola.

In the early months of 1985, the company was in crisis and its very survival seemed doubtful. The following sections highlight the difficulties and examine the options for corporate redesign in the context of the intelligent system attributes presented in Part II.

The issues relating to each dimension of organizational design are highlighted in Table 13.1 in conjunction with their modes of resolution. The entries in this table are examined in greater depth below.

## Purpose

The vision of Apple Computer was to render computing technology accessible to the non-technical user. The Apple machines would appeal to educational markets running the full gamut from kindergarten through high school and to home users for applications such as tutoring, personal finance, and game playing. As with many new consumer technologies, computers would first appeal to a small group of hobbyists and enthusiasts, then expand into larger markets represented by ordinary users.

It was Jobs' view that the home computer would become an appliance, much like a toaster or telephone. It would therefore be a stand-alone device, a closed system without the capability to communicate with other computers. If a component failed, it would be replaced rather than fixed; if the machine were superceded by a newer model, the older one would simply be discarded.

*Table 13.1* Organizational design issues and their resolution

| Dimension | Issue | Resolution |
|---|---|---|
| 1. Purpose | Developing innovative products | Increase R & D |
| | Promoting employee welfare | Encourage individual initiative; tolerate iconoclastic behavior; promote "Apple culture" |
| 2. Space | Potential of business markets | Systems approach, including AppleTalk |
| | Coexistence with IBM and other vendors | Open architecture |
| | Noncommittal software developers | Forging agreements with the external infrastructure |
| | | Changing ad agencies |
| 3. Structure | Deadlock between Sculley and Jobs | Repositioning Jobs |
| | Clash between Apple II and Macintosh divisions | Switching from divisional to functional structure |
| | Maintaining vitality | Creating "spinout" companies |
| 4. Time | Delays in product development | Increase in product development budget by 70% |
| | Leadership in technological innovation | Hiring talented engineers and encouraging "wild" ideas |
| | Future focus | Imagining the future, and working back to the past |
| 5. Process | Neutralization of middle management | Participatory management; delegation of responsibility |
| | Employee morale | "One Apple" |
| | Employee initiative | "Each person makes a difference" |
| 6. Efficiency | Trimming expenses | Laying off 20% of workforce; closing 3 out of 6 factories; Newspaper ads, not TV |
| | Maintaining profit margins | Merging Lisa into Macintosh division |
| | | Focus on upper end of product markets |

The original Apple machines had evolved into the Apple II product line, and were joined in the early 1980's by Macintosh models built on a new generation of microprocessors and software technology. The Macintosh was designed to interact with the user through a keyboard and a mouse, and provided a sophisticated graphics interface, at a price tag of $3,000–$6,000. This line was dedicated to the business and university environment. A key component of the business strategy was to allow for Apple machines to communicate through a local network called AppleTalk.

The objective of the Apple Computer Company was to become one of the most successful firms in the world. More specifically, its goals were of two types:

- *Product innovation.* The firm would generate high-quality products incorporating state-of-the art hardware and software.
- *Employee growth.* The individual would enjoy a benign, supportive culture such as that at Hewlett-Packard, and be made to feel his work would make a difference to the company.

Unfortunately, the organization satisfied neither goal in 1985. For example, the Lisa computer failed ignominiously in the marketplace: at a $10,000 price tag it was too expensive to penetrate target markets and too unsophisticated for heavy-duty tasks. In fact, the machine had to be abandoned only months after its introduction. Further, no new products were forthcoming from the Macintosh division, a group Jobs regarded as his own domain.

The cultural aspects of the company were not much better. The initial friendship between Jobs and Sculley had deteriorated into confrontation. Jobs' penchant for innovative and dynamic thinking was one source of contention; he was given to delving into organizational concerns at any level, and issuing directives that bypassed middle management. Sculley, on the other hand, advocated a more participatory mode of management, and wanted Jobs to limit his activities to product innovation.

Another problem was the ill will between the Apple and Macintosh divisions, a conflict that over time led to many resignations. Among the Macintosh personnel, Jobs cultivated an *esprit de corps* that bordered on elitism, and too often "borrowed" ideas or cancelled projects without notice in the Apple II division.

The problem of low morale, as well as the paralysis in product development in the Macintosh division, disappeared after Jobs was released from his operating role in May 1985. He still retained his other title as chairman of the board, a position of nominal significance and lacking in operating responsibility. By concentrating on the function he could best perform, Jobs was both freed from mundane administrative responsibilities and prevented from placing his personal stamp on organizational processes. In addition, the rift between the Macintosh and Apple II divisions was bridged by the conversion from a divisional to a functional structure; former rivals could again become colleagues engaged in the pursuit of a common goal.

The problems at Apple Computer stemmed, in part, from a nebulous definition of its mission and ways to realize it. First of all, trying to be all things to all people is a risky strategy, especially for a fledgling company. "A computer for every purse and purpose" might possibly work for an IBM, but not for a debutant. The new company's lack of focus was apparent in the introduction of the Lisa. This computer carried the price tag of a business machine, but not the hardware capabilities and software repertoire required by serious users.

Meanwhile, back at the company, the internal goals of product innovation and employee growth were laudable objectives; but these are goals without substance unless supported by an internal infrastructure. The misunderstanding of goals and roles among key personnel led to problems for the rank-and-file employees. As we will later see, the structure of employee incentives did nothing to promote the stated goals of the organization.

# Space

For years, Apple had been the premier developer of personal computers, but by the early 1980's it faced a large array of competitors. In response to this new field of challengers, Apple focused its efforts on the middle and upper end of the personal computer market. Most of its competitors—such as Atari, Commodore, Tandy (Radio Shack) and Texas Instruments—seemed satisfied to remain at the lower end. The market leaders in personal computers at the beginning of the decade are given in Table 13.2.

The Apple II, which had given birth to the company in 1976, served as the main product line. It was joined in 1980 by the Apple III for higher-end users such as small businesses. Jobs' dedication to an emphasis on external appearance led to internal design problems in the new machine: for instance, the wires were packed too tightly and caused short circuits. The machine was of poor production quality, and offered little in the way of application software. By 1983, Apple III sales reached $100 million; but the machine was considered a failure by industry observers.

In 1980 Apple Computer controlled 80 percent of the personal computer market. When IBM announced its entry into the personal computer fray in 1981, Apple remained nonchalant. Confident of its technical and marketing prowess, Apple ran an advertising campaign with the greeting, "Welcome IBM. Seriously." The IBM PC was technologically weaker than the Macintosh; the 8086 microprocessor used a 16-bit architecture rather the the 16/32-bit format of the M68000 in the Macintosh. A larger drawback, perhaps more condemning, was the dismal user interface. A low-end model subsequently introduced by IBM, the PCjr, was little more than a toy that flopped in the marketplace and was eventually withdrawn.

Despite the drawbacks, the sterling name of IBM, coupled with a hefty advertis-

*Table 13.2* U.S. Personal computer leaders in the market, 1982 (*Source*: *Datamation*, June 1983, p. 89.)

| Company | 1982 PC Revenues ($ Millions) | 1981 PC Revenues ($ Millions) | % Increase |
|---------|------------------------------|------------------------------|------------|
| 1. Apple | 664 | 401.1 | 65.5 |
| 2. IBM | 500 | N/A | N/A |
| 3. Tandy | 466 | 293 | 59 |
| 4. Hewlett-Packard | 235 | 195 | 20.6 |
| 5. Texas Instruments | 233 | 144.3 | 61.4 |
| 6. Digital Equipment | 200 | N/A | N/A |
|  |  | Average Growth: 51.6% | |

ing budget and an unmatched reputation for service, sufficed to attract numerous customers. After its entry into the arena, IBM garnered 18 percent of the personal computer market in 1982 and 38 percent in 1983. The high-end models, the PC/AT and PC/XT, priced at $2,500 to $6,000, also turned in respectable performances in the market. Another major factor in the market was Compaq Computer, which offered IBM-compatible machines of greater power and lower cost than the PC line. In addition, IBM and its compatible machines attracted a huge library of applications software from external developers. In a strategic marketing move, the computer giant introduced a version of Unix, the versatile operating system software, before Apple did.

Apple Computer introduced the new-generation Lisa in 1983. Although the machine enjoyed brisk sales initially, it came to be viewed as a loser. The Lisa suffered from a delayed introduction and labored under a hefty price tag. It was plagued by the same type of production problems as the Apple III and could not communicate with equipment from other vendors. Since Jobs had not been obliging to the infrastructure of accessory providers, the Lisa—like the Apple III—had no applications software.

In a drastic move, the price of the Lisa was reduced to the range of $3,500 to $4,500 depending on the model. But a price cut could not overcome other deficiencies. The new machine was abandoned within months after its introduction, after a career even briefer than that of the Apple III. The resources of the Lisa division were then merged into those of Macintosh.

Meanwhile, the Apple II performed well in the marketplace, yielding 300,000 units of sales in 1982 and 700,000 in 1983. In 1984 the Apple IIe and the IIc models enjoyed 800,000 units in sales, generating $1 billion and two thirds of corporate profits. The first half of 1983 was a halcyon period for Apple, and its stock rose to $63.75 per share. But the personal computer market collapsed in the middle of the year, bruising the golden Apple. Revenues plunged, Apple suffered its first loss in quarterly profits, and its stock plummeted to $17.24 per share.

Apple's ownership of the personal computer market fell from 21 percent in 1982 to 19 percent in 1983. In the same period, IBM's portion of the market grew from nothing in 1981 when it first entered the marketplace, to 18 percent in 1982 and 38 percent in 1983.

When Sculley arrived at Apple in 1983, his task was three-fold: to cut costs, repair relations with the personal computer infrastructure, and reposition Apple in the marketplace. The first task relates primarily to organizational structure, and is explored more fully in the next section. The second objective was fulfilled by a comprehensive program of speeches and personal discussions between Apple executives and key suppliers, dealers, customers, and market analysts.

The repositioning of the company was achieved in part by providing for an open architecture to permit interfacing with machines from other vendors. Another concern was the confusion in the marketplace over Apple's role in the computer world. Was the company a low-end supplier to hobbyists, households, and grade schools, as embodied by the Apple II, or a small business vendor symbolized by the Apple III? A professional equipment producer as in the Lisa, or business and university suppliers embodied by the forthcoming Macintosh? To create a unified theme and a focused position in the consumer's mind, the company changed advertising agencies and adopted the slogan, "The power to be your best."

Instead of confronting IBM directly, Apple opted for coexistence. This strategy was implemented by providing for accessories to permit communication with IBM devices.

The Macintosh was finally introduced in January 1984. A key marketing advantage of this machine was its capacity for "desk-top" publishing. The ability to intermix graphics and text with ease on a single page, coupled with the crispness of output on the LaserWriter printer, priced at $6,000, permitted even small offices to produce professional-looking reports and brochures at a fraction of the cost of contracting the work to commercial agencies.

Geographically, Apple had an international presence that was especially strong in France. The company operated manufacturing facilities worldwide: a highly automated facility in Fremont, California, for producing Macintoshes; a plant in Garden Grove, California, for making keyboards and the mouse; a productive but labor-intensive factory in Carrolltown, Texas, for the Apple IIc computer; plants for making computer accessories at Mill Street and Cork, Ireland, and a components factory in Singapore.

An organization cannot afford to ignore its environment. Despite its success in the household and educational markets, Apple Computer could not have expected to take the business market by storm against IBM. Loyalty among previous customers and respect among potential clients for the computer giant often weighed a decision in IBM's favor, even when Apple's products were technically superior.

Apple Computer also found it imperative to establish ties with key sources in its environment, namely, software houses, suppliers, dealers, and market analysts. Most important, the company had to respond to the needs of its lifeline, the customers. A milestone was passed in this direction with the adoption of an open architecture. As a result of the change, customers could safely purchase an Apple machine without fear of having to discard old equipment with growing needs, or of imprisonment to a single vendor's hardware.

## Structure

In 1983, Apple Computer had a divisional structure based on different product lines. The Apple II was the staple product line, feeding the development of the others: Lisa, the high-end professional machine; Macintosh, the innovative low-end office machine; and Apple III, which would bridge the gap between the Apple II and Lisa. Further, auxiliary products such as peripheral devices and accessories were organized into separate divisions. This structure led to the duplication of effort as well as intense rivalry.

There was also a peculiar relationship between Sculley and Jobs. Sculley was the president and chief executive, but Jobs was the chairman of the board and general manager of the Macintosh division. As chairman, Jobs was superior to the president, but as general manager he was subordinate. Difficulties arose from Jobs' decisive temperament and personal convictions; he was a visionary but less experienced as a manager.

Part of the problem at Apple was lack of focused leadership. Jobs had delved personally into other employees' responsibilities. Sculley, being a newcomer to the corporation, wanted to avoid a confrontation. Although Sculley was experienced in the ways of corporate America, he was new to the computer industry. Most of the other employees at the company were new to the business world.

A time of crisis, however, is a time for strong leadership and centralization. With the support of Markkula and the board of directors, Sculley released Jobs from his operating function; the latter retained his role as chairman and was to serve as a guiding influence on the company's long-term vision. The divisional structure had served Apple poorly, as each product division possessed its own sales, marketing, development, and manufacturing groups. Even worse than the attendant redundancy in jobs was the rivalry among differing divisions; the decentralization had prevented cooperation and synergism.

The divisional organization was scrapped in favor of functional lines. Delbert Yocum, general manager of the Apple II division, was appointed group executive in charge of all operations, ranging from product development to production and logistics. Reporting to him was a vice president in charge of all product development, and another dedicated to worldwide manufacturing operations. Of equal significance were the merging of U.S. marketing and sales under one executive, and of the corresponding international activities under another. A chief financial officer as well as a human resource executive were added to the top-level staff.

This structure succeeded admirably in carrying the corporation through its stormy days. Over time, the organizational structure would continue to undergo minor changes to better serve the needs of the growing company in a turbulent environment.

Although Apple continued to expand, the management wanted to foster a small-company culture. To do this, Apple encouraged the formation of new business ventures called "spinouts." The first of these offspring, Apple Software, was conceived in 1987 with the mission of creating software packages for Apple machines. After a year of nurturing, the young organization was christened Clovis, granted independence, and released into the marketplace to seek its future.

Unlike the conventional spinoff company that pursues its own destiny independently of its parent organization, a spinout is intended to maintain close business ties. By developing a network of spinouts, the parent company not only maintains its lean structure but creates a confederation of allies, a "negotiated environment" of a special kind.

The structure of an organization should be directed toward its objectives. The dual role of Jobs as chairman of the board and general manager, coupled with his strong personal style, hindered rather than augmented Sculley's efforts as president. Only when this convoluted structure was simplified could a rational managerial policy be implemented.

Further, the divisional structure of the company was inappropriate in light of organizational capabilities. A number of older, established companies may enjoy the good fortune of having redundancy in talent: several competent employees to fill each critical role in line and staff positions. A young, growing company such as Apple, however, does not enjoy this luxury. The divisional structure only served to spread out a scarce talent base and to incite rivalry. The new functional structure was more in line with organizational resources, and also facilitated the coordination and implementation of business strategy.

Finally, the creation of spinout companies allowed for the replication of entrepreneurial culture in additional organizations and the reaffirmation of small-company spirit within Apple. The spinouts also served as sources of support in the industrial environment, thereby creating a special alliance with external groups and enhancing the stability of the parent company.

# Time

A key difficulty faced by the company in 1984 was the lack of innovation, which had become especially acute within the Macintosh division. The first Macintosh model was not sufficiently powerful to appeal widely to business or professional users; the later, more powerful Lisa—alias XL—failed in the marketplace due to its excessive price tag. Meanwhile, no new machines were imminent.

Despite severe financial pressures, top management recognized the need to deliver a healthy stream of products. While outsiders wondered whether Apple would earn profits—or even survive—in the coming year, the corporation increased product development expenses by 70 percent. Part of the investment went toward a Cray supercomputer, purchased for $15 million in January 1986 to perform design simulations and accelerate the delivery of software development tools.[2]

By 1987 Apple was investing $185 million in research and development, as compared to $40 million four years earlier. In an industry where three years separate one product generation from the next, approximately 20 percent of the investment in research was dedicated to ideas whose fruition lay two or three years away.[3]

The foundations of technological leadership are found in the fertile minds of creative individuals. As with many other companies at the cutting edge of computer technology, Apple sought to promote achievement by encouraging risk-taking while relaxing controls.

To prepare the company for the future, Apple employees spend part of their time dreaming about it, then backtracking to the present to determine how best to realize the dream. A case in point is the Knowledge Navigator, a software package which is imagined to be available in the next century. A person would employ the Navigator as a vehicle to access an unbounded store of databases, libraries, museums, and other sources of knowledge, all the while interfacing with the system through the senses of sound, touch, and vision.

A forerunner of the Navigator was implemented in the form of HyperCard, a set of software tools for organizing knowledge in novel ways. The user can establish links among diverse items of information, then access them in flexible ways by merely pointing at words or pictures of index cards shown on a video screen. This package and its applications were designed as a stepping-stone to more advanced systems in the future.

To ensure its long-term viability, an intelligent system—whether a machine, an organization, or an entire society—must deploy its resources wisely. No foresight is required to funnel resources into short-term needs, but a prudent system must also take long-term requirements into account. To this end, Apple invested heavily in research and development even at a time when its immediate needs might easily have absorbed all its financial capital as well as the time and energy of the managerial and technical leadership.

## Process

In the top management's view, the business market represented the major opportunity for future corporate growth. This market was to be served by the Macintosh product line, in contrast to the Apple II series designed for the home and lower educational segments.

The emphasis on Macintosh strained internal relations, as both Jobs and Sculley seemed to favor the new Macintosh at a time when its development was largely funded by Apple II revenues. Although slighted by Jobs and Sculley, the Apple II group set performance records in 1984. The sale of 800,000 units of the IIe and portable IIc models brought in almost $1 billion in revenues and accounted for two-thirds of the company's profits.[4]

In the early 1980's, the middle management at Apple was becoming demoralized by Jobs' strong leadership style. A case in point is the story behind Macintosh. This product was conceived in 1979 by Jef Raskin as a machine for the business market. Jobs, who was personally committed to developing tools for individuals rather than organizations, disagreed with the concept. Despite the chairman's opposition, Raskin garnered enough support from the executive board to continue developing the machine. Over the course of time, however, the promise of the Macintosh became more apparent. Jobs began to exert influence over the division in 1980, and took over completely in 1982 after the departure of a disenchanted Raskin.

In an attempt to build morale and enhance productivity, Jobs showered his newly-adopted division with attention, resources, and fringe benefits. These actions, however, were perceived by other employees as an unwarranted display of favoritism at the expense of the other groups. The rivalry between Macintosh and Apple II intensified in 1984, and the company suffered a steady stream of resignations.

Although Wozniak used his influence to try to improve relations, his distaste for politics limited his efforts. The problem of low morale was addressed in 1985 by relieving Jobs of operating responsibility, followed by the delegation of responsibility and authority to middle management.

To further palliate the rivalry and bitterness, a campaign was launched under the banner of "One Apple," designed as much to reassure external observers as employees. Employee initiative was nurtured by emphasizing the uniqueness of the Apple culture and the conviction that each person can make a difference.

The processes in an intelligent system must be designed to fullfill its objectives.

A major objective of Apple was the attainment of employee harmony. However, Jobs's strong style did not endear him to many subordinates. This problem was exacerbated by the apparent emphasis of both Jobs and Sculley on the new Macintosh group at a time when its development, as well as the company as a whole, was supported largely by Apple II revenues.

The first task of management was to recognize the disparity between objectives and implementation. This was followed by a modification of internal processes to provide a better balance in engaging different product groups, including a new program to reaffirm the unity of the Apple culture as well as respect for individual contributions.

## Efficiency

With the collapse of the personal computer market in 1983, Apple Computer stipulated a hiring freeze and announced that no dividends would be paid in 1984. To further slash expenses, three out of the six factories were closed in 1985. The Mill Street plant in Ireland was merged with the one in Cork; the Garden Grove plant in California integrated into the Fremont plant; and the Carrolltown factory in Texas shut down. In the process 1,200 people, representing 20 percent of the workforce, were laid off; many of these were from the manufacturing and marketing divisions.

The advertising budget was slashed from $100 million in 1984 down to $45 million in 1985. Apple's ad agency was changed, while newsprint advertising replaced expensive television promotions. All the while, profit margins were maintained by focusing on the upper end of the personal computer market, including business users.

An intelligent system cannot afford to pursue its goals in haphazard fashion. Its very survival may be jeopardized by wasting resources through duplicated effort or ineffectual strategies.

Precipitated by the crisis of 1983, a number of austerity measures were implemented at Apple Computer. These measures included reductions in both fixed expenses, such as the closure of factories, and variable costs, such as dividends and advertising. By trimming its excess weight, Apple Computer regained its strength and was poised for a new assault on the markets.

## Results

When Jobs introduced the Macintosh in January 1984, it was promptly criticized for anemic performance: according to the detractors, much of the power of the 16/32-bit architecture was directed toward graphic manipulation rather than "serious" computation. Further, the Macintosh's closed architecture permitted no hookup with machines from other vendors.

The initial Macintosh, supplied with 128K characters of memory, failed to penetrate the business market, but it was quickly superceded by models with larger memories: the 256K, then the 512K, followed by the Macintosh Plus. The Plus, a machine offering 1024K, was introduced in 1986 and made some inroads in the business world. The company sold 174,000 Macintoshes in 1984; 245,000 in 1985; and 480,000 in 1986.[5] Subsequently the Macintosh SE with a 1024K memory, dual diskette drives, and an open configuration was introduced, in conjunction with its sibling Macintosh II, offering a 20 Megabyte internal hard disk and a single diskette drive. These were widely welcomed as serious machines appropriate for commercial and professional applications.

During the darkest days of 1985, Apple Computer established an ambitious multi-attribute goal for the coming year: 50 percent gross margin on products shipped, 50,000 Macintoshes sold per month, and a stock price of $50 per share. The first goal was reached by the end of 1985, and the latter two by early 1987.

The corporation earned an income of $154 million on net sales of $1.9 billion in 1986. The combined value of Apple stock soared from $900 million in June 1985 to $6.0 billion in spring 1990.

According to a caveat in the investment industry, however, "past performance is no guarantee of future success." Time will tell whether Apple Computer continues to flourish in the future as it has so ably done in the past. As with all organizations, Apple Computer will have to evolve on a continuing basis as it moves into the new millennium. Its task will be all the more challenging, as it faces an environment characterized by increasing levels of turbulence and technological discontinuity.

## Summary

This chapter has examined the internal and external forces acting on Apple Computer, Inc., during a critical stage of its development. The interplay of these factors, and the attendant decisions concerning organizational structures and processes, can be understood in terms of the general framework for intelligent systems. The issues and trade-offs involved are similar for the design of intelligent systems, whether they are largely physical as in the case of a robot, or primarily abstract as in an organization.

# V

# POSTLUDE

*History, science and philosophy all make us aware of the great collective achievements of mankind. It would be well if every civilized human being had a sense of these achievements and a realization of the possibility of greater things to come, with the indifference which must result as regards the petty squabbles upon which the passions of individuals and nations are wastefully squandered.*[1]

Bertrand Russell

# 14

# Conclusion

*It will be the business of reason to rule with wisdom and forethought on behalf of the entire soul.*[1]

Plato

The social scientist differentiates between a theory of what *is* versus what *should be*. More specifically, a *positive* model or theory simply describes a phenomenon, while a *normative* one prescribes a set of principles of behavior. For example, a positive theory of human action may state as an axiom, "People are greedy"; but such a tenet is unlikely to comprise part of a normative theory.

In the physical and life sciences, the distinction between descriptive and prescriptive models may have been relevant in previous eras, before the rise of the scientific method. This was especially true when teleological interpretations were superimposed on the behavior of inanimate objects, even in situations where such imputations would be considered unnecessary or even inappropriate in modern practice.

For example, the fact that a stone falls to the ground was interpreted as the result of the following principles: first, the "natural" place for a stone is its resting state, which is the ground; and second, a stone desires to attain its proper place in the universe. In a similar way, it was proper that a flame, being light, should ascend into the heavens.

The advent of the scientific method brought analysis to the forefront of investigations into natural phenomena. Since analysis deals with the *what* or *how* of things rather than the *why*, teleological interpretations were swept aside—and with them the prescriptive perspective.

The analysis of intelligent systems is largely a descriptive effort, as exemplified by techniques for measuring the information content of communication signals. The synthesis of such systems, however, has a prescriptive component. A theory of intelligent systems, to be useful for the design phase, must have normative rules to guide engineering activity.

This prescriptive aspect may be explicit, as in the rule "Minimize complexity," or it may be implicit, as in "A good design is one that has minimal complexity." But the prescriptive ingredient, however faint, should flavor the theory.

A full-fledged theory of intelligent systems is yet to come. To that end, the current book has taken one step by proposing a general framework to serve as a unifying structure for approaching the analytic and synthetic issues in a systematic way. Further, this framework may serve as the foundation for a more comprehensive theory that should emerge in the years ahead.

## Networks of Intelligence

From an operational standpoint, observable behavior is the only basis for attributing intelligence to a system. According to one view, intelligent behavior is in fact the telling factor in discriminating between life and non-life.[2] From this perspective, the intelligent structures that we design and build today may be the forerunners of an alternate form of life, which, some believe, will eventually outlive their creators.[3] Others might merely note that the follies of mankind will ensure that neither form will endure.

Even if our species were to vanish from the planet, however, our influence should persist. A world inhabited by the descendants of our creative efforts will likely embody our conceptions of the universe.[4]

Perhaps a more probable scenario is the symbiosis of intelligence and alternative life forms. The human and machine, in particular, will develop a more intimate relationship over the course of time. The human will become more mechanical, through developments such as intelligent prosthetics and implanted devices for telecommunication. In parallel, certain intelligent devices will become more life-like. These will include general-purpose robots which may take anthropomorphic form, replete with sensory organs, supple hands, and conversational capabilities.

In this way, the distinction between the natural and artificial will become more diffuse. Someday the human and machine may constitute the extremes of a continuum rather than a dichotomy.

Should current trends continue, much of our routine environment will become animated. Our appliances, buildings, vehicles, and roadways are already incorporating intelligence as well as the ability to communicate with other agents. Hence each entity, whether a person, machine, or hybrid, will become part of a larger network of intelligence.

Such developments would bring with them a myriad of concerns. An obvious issue is that of privacy. In large cities today, apartment dwellers are less likely to know their neighbors than those who live in small towns. When a person is vulnerable to constant monitoring, she may guard her solitude more preciously than otherwise.

Hence the society may need to develop mechanisms to provide some measure of privacy. The extent to which such concerns are addressed, and the vehicles for implementing the decisions, will surely generate much debate and even passion in the years ahead.

A discussion of positive and normative models of human activity has relevance not only to the design of intelligent systems, but to the uses to which they may be commissioned. Our expanding capacity to remodel the environment implies a need to direct our activities with prudence. We possess the power to create intelligent devices and refashion our institutions as well as to alter the landscape around us. We can effect these changes for the better or for worse.

## *Opportunities*

Our challenge in the decades ahead is to design a flexible, robust society. The future that we fashion should be resilient, forgiving of mistakes that will surely result from our incomplete understanding of the physical world as well as from human nature and societal dynamics. Any decisions we make and actions we undertake should be reversible. The suggestion has been made, for example, that our next Dark Age— following perhaps some natural calamity or a nuclear holocaust—may be permanent. If the instruments of civilization such as books, computers, and power plants were to be lost, the survivors may be left without tools to amplify their muscle or mind. And the remaining minerals as well as fossil fuels may be entrenched too deeply in the earth's crust to allow for manual excavation.

A resilient society cannot result from a single monolithic structure, but only from a decentralized one fostering diversity and redundancy. There will be only one future, but that future must allow for diversity.[5] In the words of John Fitzgerald Kennedy, "The wave of the future is not the conquest of the world by a single dogmatic creed but the liberation of the diverse energies of free nations and free men."[6]

Hence the increase in our range of capabilities, amplified by the instruments of intelligent systems, underscores the need for a rational theory of human action. The effort to develop a rational policy toward intelligent systems will draw on normative theories as well as prescriptive principles. In this task our past experience in both the natural and social sciences will serve as a guide.

It has often been said that we are creating a sentience that will transcend our own. Some of us would call this ironic, others fitting, and still others disturbing. Whatever our reactions may be, it appears inevitable. We will build intelligent systems of increasing sophistication and in turn be shaped by them. This destiny lies in our stars since the creative drive springs from within.

# APPENDICES

*If the possession of information can be understood as the posses-
sion of some internal physical order that bears some systematic
relation to the environment, then the operations of intelligence,
abstractly conceived, turn out to be just a high-grade version of
the operations characteristic of life, save that they are even more
intricately coupled to the environment.*[1]

Paul Churchland

*Our genes may be immortal but the collection of genes which is
any one of us is bound to crumble away. Elizabeth II is a direct
descendant of William the Conquerer. Yet it is quite probable that
she bears not a single one of the old king's genes. We should not
seek immortality in reproduction. But if you contribute to the
world's culture, if you have a good idea, compose a tune, invent a
sparking plug, write a poem, it may live on, intact, long after
your genes have dissolved in the common pool.*[2]

Richard Dawkins

## Introduction to Appendices

A science of intelligent systems must be based on a general framework that encompasses the salient features of smart behavior and principles for designing sophisticated entities. The framework must support rigorous models through the use of
mathematical tools such as symbolic logic and quantitative metrics.

One direction for the future lies in the incorporation of these constructs within a
general framework for design. A general model of design as a sequence of computational transitions is discussed in Appendix A.

The reasoning behind intelligent behavoir can be viewed in terms of the formal
models of thought as embodied in mathematical logic. The power and limitations of
reasoning and creative thinking, from the perspective of logical methods, is examined in Appendix B.

The general model of design given in Appendix A can accommodate principles for design such as those for intelligent systems, as well as other aproaches such as the axiomatic approach. To date, the axiomatic methodology presented in Appendix C has been applied primarily to the synthesis of conventional rather than smart systems. Since the design axioms, in essence, call for simplicity in design, its philosophy must apply universally. Continuing research will indicate how the design axioms may best be interpreted in the context of smart artifacts.

A useful framework for intelligence should be compatible upward to generalized models of design (see Appendix A) and accommodating downward for subsets of smart behavior such as autonomous learning. It should encompass formal techniques to explore the nature and limits of intelligent behavior (Appendix B). It should interface gracefully with various principles for design (Appendix C), accommodate formal models of intelligent behavior (Appendix D), as well as metrics for performance evaluation (Appendix E).

Finally, an intelligent framework should also facilitate its own development. This may be achieved by exploring tools for employing the framework against its own attributes, as illustrated in Appendix F.

The six factors of intelligence—purpose, space, structure, time, process, and efficiency—satisfy these objectives. The appendices explore a number of these critical issues, and provide a glimpse of the road ahead in the systematic exploration of intelligence and its implementation in engineered systems.

# APPENDIX A

# A General Model
# of Design

*The proper study of mankind is the science of design, not only as the
professional component of a technical education but as a core discipline for every liberally educated person.* [1]

Herbert Alexander Simon

Design is a pervasive human activity inherent in enterprises ranging from the
construction of an engineered device to the formulation of a corporate strategy and
the organization of a social community. To the extent that commonalities underlie
these activities, it should be possible to (a) identify a generic framework of concepts
and principles, (b) organize them with sufficient clarity for translation into a formal
model, and (c) encode them into a software package to assist in decision making.
This appendix discusses the underlying premises of this approach, develops a formal framework, and discusses a system architecture for such a generic design
advisor. The presentation is adapted from a previous article by the author.[2]

## Design Process

### Framework for Types of Design

Design tasks may be categorized into a two-dimensional framework organized
according to focus and originality, as shown in Figure A.1. The concept of *focus* or
*deliberation* relates to the level of anticipation or degree to which the functional
requirements are known in advance. If the goals are known *a priori*, then we may
call the design task *focused* or *closed;* otherwise, the task is called *unfocused* or
*open.*

Most design activity is of the focused type. For example, a new factory will
have production requirements that must be fulfilled by a particular implementation.
On the other hand, an example of open exploration lies in the invention of a
patentable three-dimensional electronic component, a feat accomplished by a software package called Eurisko through interaction with its creator, Douglas Lenat.[3]

*Originality* relates to the novelty of the implementation. If the end result of a
design activity is already part of the stock of human knowledge, we may call it a

**Focus**

| | Focused | Unfocused |
|---|---|---|
| **Self-learning** | Intended conventional product.<br><br>E.g. Redesign of engine | Open-ended "rediscovery".<br><br>E.g. AM program |
| **Original** | Intended novelty.<br><br>E.g. New production process | Open-ended discovery.<br><br>E.g. Eurisko program |

*(Vertical axis label: Originality)*

**Figure A.1** A framework for types of design activity.

*conventional* implementation, and the associated process of discovery a *self-learning* phenomenon. A design that adds to the human knowledge base, on the other hand, is a *novel* product, and the process *original*.

To illustrate, the design of a portfolio of capital market investments using existing stocks is a conventional result, while the conception of a new interest-rate swap or hitherto unknown tax strategy would be an original creation. A well-known example of an open-ended, self-learning program is AM, the automated mathematician.[4]

## Components of a Design Task

The design process begins with a perceived need or problem delineated in terms of a set of specifications consisting of functional requirements and constraints which must be mapped into a design (see Figure A.2). The design is itself characterized by objects and parameters. These concepts are described in greater detail below.

The set of *functional requirements* $\mathbf{F} = \{F_1, F_2, \ldots, F_m\}$ constitutes the minimal set of independent specifications that particularizes the problem at hand. The specifications are independent in two senses: (1) they are not redundant, and (2) each specification must be satisfied to solve the problem. Each requirement $F_i$ is usually associated with a tolerance band consisting of a lower bound $F_{i0}$ and an upper bound $F_{i1}$ . In other words, the $i$th specification must be satisfied by a design that operates within the closed interval $[F_{i0}, F_{i1}]$. For example, the thickness of a slab of polymer might be specified as 2 cm, plus or minus 1 mm.

Often design problems are characterized by a set **C** of *constraints* that further limit the bounds of acceptable solutions. For example, a functional requirement of a mobile robot may indicate that the machine detect noise, while a constraint might stipulate the robot frame should fit through a standard 30-inch doorway, or cost less than $2,000.

The goal of the design task is to identify an *implementation* or *design* **d** that meets the specifications. The implementation may be a tangible thing, such as a glider, or intangible, such as a software package or an organizational structure.

The design is defined by a set of *objects* or *features* $\mathbf{f} = \{f_1, f_2, \ldots, f_q\}$ that serves as components of the implementation. The features are in turn characterized by a set of *parameters* $\mathbf{p} = \{p_1, p_2, \ldots, p_r\}$. The parameters qualify the features in various ways such as size, weight, color, etc. These parameters are the critical

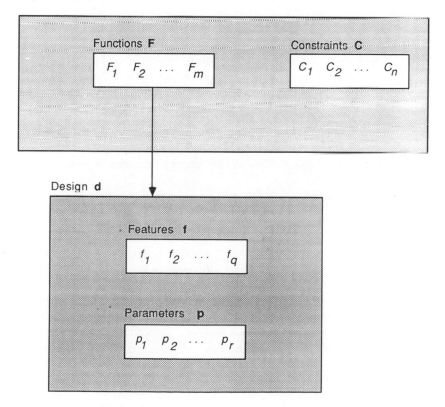

**Figure A.2** Basic components of a design task.

attributes of a design that relate to the functional requirements **F**. Each object or feature may have one or more parameters.

To illustrate, consider an automotive engine. A key functional requirement is to deliver a specified level of power, while abiding by constraints such as weight and size. The engine block and pistons are examples of design features or objects. For a fixed depth of piston traversal, the diameter of the pistons is a parameter that may serve to satisfy the desired power output.

## *Transitions in Design*

As stated above, the design process involves a mapping of functional requirements into a set of features and parameters. Only for trivial problems can this mapping be accomplished in a single step: in general, design is accomplished in a series of *transitions* or *transformations* from one state of knowledge to another.

More specifically, design activity may be viewed as a sequence $\langle d_0, d_1, d_2, \ldots, d_z \rangle$ which involves the evolution from a null design $d_0$ to the final design $d_z$. The transitions occur as a result of a set **T** of *transformation* operators (see Figure A.3).

At the beginning of the design task, the design $d_0$ consists of a null description (no features or parameters). In response to the specifications, a preliminary description $d_1$ is generated, then checked against the specifications using a set of *criteria* **Cr**. If the specifications are not met, a transformation $T$ is performed on $d_1$ in order to yield the next iteration $d_2$. At each iteration $j$, the selection of a specific transformation operator depends on both the functions **F** and the partial design history $\langle d_0, d_1, \ldots, d_{j-1} \rangle$. The process continues until one of the following conditions is met:

1. The specifications are satisfied.
2. The impossibility of attaining the specifications, given the available knowledge base, is recognized.
3. Continuing the design process can no longer be justified in terms of expected benefits and costs.

The perspective of design as a transformational activity may be further categorized into sequential versus parallel modes. The sequential viewpoint, also called the abstract refinement model, involves the transformation of a set of goals into a hierarchical, top-down strategy in search of the data or initial facts.[5] In other words, each step involves the replacement of a single node in the tree of descriptions by a more detailed description.

The parallel viewpoint involves the transformation of a functional requirement into an implementation, while operating on one or more components simultaneously.[6] The conversions involve transitions from one complete description to another. Since the parallel viewpoint can accommodate more than one component at a time, it is more general than the abstract refinement model. But the price for generality is increased complexity.

A design system may be represented by a "black box" that acts on input specifications and transforms them into some resulting implementation. This simplified representation of the design system as a machine that operates on a series of inputs

Knowledge **K**

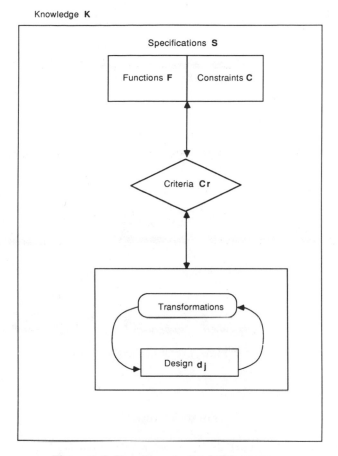

**Figure A.3** Transitions in the design process.

and produces an output, coincides with the viewpoint of Automata Theory, a correspondence which will be explored in subsequent sections.

## Transitional Model as a Metalevel Reasoning System

The transitional model may serve as a metalevel reasoning system for design, by incorporating domain knowledge to guide the design process in specific realms of application. A convenient way to organize the knowledge base is in the form of a hierarchy; an illustrative tree of application domains is depicted in Figure A.4.

In summary, the transitional model of design serves the following goals: (1) to identify generic concepts and processes inherent in design across diverse realms of application, (2) to provide the basis for a rigorous formal theory of design, and (3) to serve as a metalevel reasoning structure for a knowledge-based system to assist in design. The first objective was pursued in this section; the second two goals are explored below.

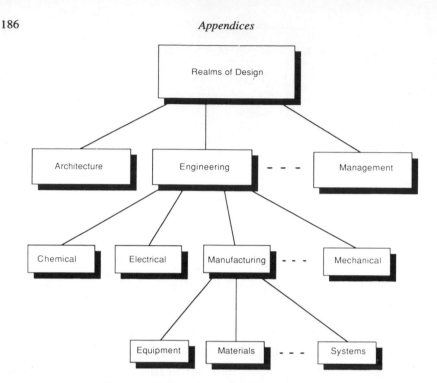

**Figure A.4** Partial tree of application domains.


## Formal Models

A formal model of design should be representable ultimately by the fundamental principles of mathematics, relating to set theory and symbolic logic.[7] By building on this foundation, it also becomes possible to draw on the work in fields such as Computability Theory and Automata Theory to assist in the development of a formal theory of design.

The isomorphism among the implementational, computational and formal realms of design is illustrated in Figure A.5. In the implementational realm of design, the design process takes a set of specifications and derives the desired final product. Similarly, in programming, the facts describing the initial state are manipulated by a set of procedures to compute the final value. In symbolic logic, the axioms and a set of proof or inference procedures are used to generate theorems or conclusions.


### *Logical Representation*

From the perspective of mathematical logic, the functional requirements and knowledge of the application domain may be included as a set **A** of initial statements or *axioms*. The set **A** also includes statements that define the criteria **Cr** for acceptabil-

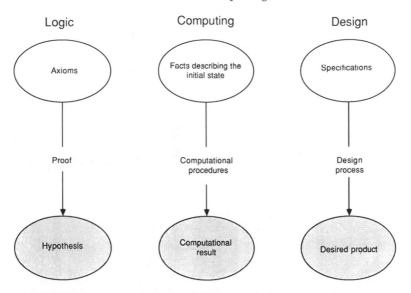

Logic      Computing      Design

**Figure A.5** Isomorphism among the formal, computational, and design realms.

ity of the final design. In other words, the criteria define the mapping from object level attributes to the functional requirements.

The initial set **A** may be modified into a sequence of related sets $\langle \mathbf{A}, \mathbf{A}_1, \mathbf{A}_2, \ldots \rangle$ using a set **T** of transformation operators. (An example of a transformation operator is *modus ponens*, which states that if the two facts "*P* implies *Q*" and "*P*" are known, then "*Q*" may be concluded.)

The union of the transformations, plus the evolving fact base, constitutes the *knowledge base* **K** for the design problem. In other words,

$$\mathbf{K} = \mathbf{T} \cup \mathbf{A} \cup \mathbf{A}_1 \cup \mathbf{A}_2 \cup \ldots .$$

Hence the knowledge base **K** evolves over the course of the design.

The goal of the design process is to identify some set **D** of statements that define the required design **d**.

## Automata Theoretic Representation

Engineering processes have been modeled as abstract systems that alternate among a discrete set of states.[8] The design process may also be modeled as an automaton for these reasons:

1. Automata Theory is applicable to any system that processes information.
2. The theory recognizes the discrete nature of objects.
3. The theory recognizes programming as a central topic.

Automata Theory may be defined as the study of the fundamental principles of systems that alternate among a discrete set of states, the principles being related to the structure, organization, and programming of such systems. A finite automaton consists of a finite set of states and a set of transitions from state to state.[9] Transitions occur on input symbols chosen from a set of input symbols or specifications.

A finite automaton that allows zero, one, or more transitions from a state based on the same input symbol is called a nondeterministic finite automaton. This type of device may be defined as an abstract machine that reads an input of a string of symbols and decides whether to accept or to reject the string by changing from one state to another based on the input. A nondeterministic finite automaton may be represented by a quintuple $\langle Q,S,D,I,F \rangle$ where $Q$ is the set of states, $S$ the set of symbols, $I$ the initial state, $F$ the set of final states, and $D$ the mapping from $Q \times S$ to $Q$.

From the above definitions we can see that Automata Theory is relevant to any system that transforms or changes from one state to another. Automata Theory may, therefore, serve as a cornerstone in the development of a formal model of the design process.

The Halting Theorem represents perhaps the central concept of Automata Theory: in general, no program can determine whether another program will ever halt. (The decision whether any *particular* program halts or not, is often possible.) This result will be used in the next section to derive more specific results relating to design activities.

The correspondence among the concepts in four fields—logic, computability theory, programming, and the object world of design—is depicted in Figure A.6.[10] This isomorphism supports the view that any formal model of a design system should be based on mathematical foundations and may build on the results of Automata or Computability Theory.

## Theoretical Implications

### *Limitations to Design Automation*

Formal models of a complex system can be used to determine the latter's fundamental character and limitations. Some previous work in this direction includes the following items:[11]

- Existence of theoretical limits to computerization in manufacturing.
- Inability of an automated factory to fabricate arbitrary products using well-defined processes.
- Inability of a computer system to deduce the purpose of a product or process given its specifications.

These results build on work in Computability Theory such as the Halting Theorem. The third result, for example, may be adapted to the realm of design as follows.[12]

> **Theorem**: No program can determine, in general, the purpose or function of a design given its specifications and/or its operating plan.

The reasoning behind this result is as follows. Suppose that a program $P$ exists that can deduce the purpose of a design **d** based solely on its specifications **S** and the

| CONCEPT | LOGIC | COMPUTABILITY THEORY | PROGRAMMING | DESIGN |
|---|---|---|---|---|
| **BASIC OBJECT** | Axiom | Basic Sequence | System Command | Descriptor or Feature |
| **BASIC TRANSITION** | Logical Inference | Compting Operation | Interpretation and Computation | Transformation |
| **STORAGE** | Infinite Workspace | Tape | Memory | Environment of a system |
| **RESULT** | Theorem | Turing Machine (TM) | Program | Product, e.g. car |
| **TRANSITIONAL REPRESENTATION** | Directed graph or list | Regular set of Sequences | Flowchart or Pseudocode | Design Sequence |

| CONCEPT | LOGIC | COMPUTABILITY THEORY | PROGRAMMING | DESIGN |
|---|---|---|---|---|
| **EFFECTIVE PROCEDURE** | Proof | A procedure that can be carried out by a simple machine | Algorithm | Design procedure |
| **EQUIVALENCE** | Same result for proof procedures | Same behavior or output | Same output | Same final design |
| **COMPUTABILITY** | A theorem can be proved using axioms and inference procedures | A function f(x) can be computed by a TM | A function (e.g. cube root) can be computed by a program | A feature/output can be produced by a design system |
| **ENUMERABILITY** | Members of a set can be matched one to one with integers | Number of basic sequences in the Regular Set are countable | Number of steps or calls in a program are countable | The basic transformations from the initial to the final state are countable |
| **DECIDABILITY** | Possible to determine whether any proposition P is true or false | Possible to determine whether an interpretation I exists | Value of the function can be determined | Possible to determine whether a design D can be produced |
| **DIVERGENCE** | Inconsistency: both P and not P can be derived | A TM produces two distinct outputs for the same input | Program produces two distinct results for the same input | Design system produces two distinct designs |

**Figure A.6** Isomorphism of concepts.

description $\Omega$ of its operating characteristics. The output of program $P$ will be a nonempty set $\Pi$. Assume that $A$ is one of the purposes of **d** as specified in $\Pi$. What happens when this question is posed to the program $P$: "Does operation $\Omega$, dedicated to the implementation of function $A$, ever terminate?" Since the program $P$ is supposed to know the purpose of **d** given **S** and $\Omega$, it should be able to answer the question. But this would contradict the Halting Theorem. So we conclude that program $P$ cannot exist.

The theorem stated above sheds some light on the ability of a program to fully analyze an object only in restricted domains of application. The theorem reflects the general limitations to introspection. Deducing the purpose of an arbitrary object implies an introspective component, since a system may be given a description and operating plan of itself as the input.

The preceding result may be readily extended to a series of related corollaries and theorems.

> **Corollary**: In general, given a description of a design system and its configuration, no program can determine the functions of the objects that are produced by that system.

> **Corollary**: In general, no program can deduce the overall objectives of a design system given a description of the system's architecture and activities.

> **Corollary**: A program, in general, cannot determine beforehand whether a design system will execute a specific action or will pass through a specific state given its description and the set of inputs.

The next subsection addresses the characteristics of discovery in the design process.

## Implications for Discovery, Learning, and Creativity

As discussed in the previous section, the steps from the initial state of the knowledge base to the final state (including the endpoint of the design process) may be viewed as a sequence of transformations. The transitional model of design given in Figure A.3 suggests the importance of having effective transformation operators. General techniques have been investigated for guiding the transformation process effectively. One such technique is the *novel combination method,* which uses knowledge of existing products to generate new devices.[13] Features of old devices which relate to desired functional requirements are identified and then combined to form the new product.

The novel combination method is appropriate for goal-driven discovery, in which the functional requirements, as well as potential features to satisfy those ends, are known in advance. It has been applied, for example, to the domain of fasteners such as screws.

The designer often faces tasks in which the functional requirements have no obvious solution space: it is unclear what set of features could possibly fulfill the requirements. A proposed approach to such tasks is the axiomatic design methodology, based on the simplification of features and their interrelationships. The two high-level decision rules, called axioms, call for an uncoupled set of features and the minimization of their information requirements.

For the case of creative design, it is more difficult to guide the design process. A sequential, brute-force exploration through the search space is computationally infeasible, hence the design process must be directed in some non-random fashion.

One approach is to develop heuristics that guide the exploration using criteria of "interestingness" that are domain-independent. Illustrative criteria relate to the quest for symmetry, specialization, or generalization.[14] An example of a heuristic to generalize a decision rule is to eliminate one or more conditions in the **IF** part of a clause.

A more effective strategy would be to couple such domain-independent heuristics with domain-specific rules. These concerns are discussed further in the next section.

The premise of this approach is that principles such as the foregoing can be identified, distilled into a uniform format, and formalized with logical precision. Once the principles are formalized, they can be readily transformed into software and embedded in an advisory expert system.[15] Moreover, the unifying framework provides a vehicle for ready expansion or adaptation of the software advisor.

One question that comes to mind is the degree to which an automated system can create novel concepts or objects. From a theoretical perspective, the knowledge base is a closed, predetermined system in two senses: (1) a finite knowledge base can only lead to a finite set of distinct consequences, and (2) the design process cannot uncover any new predicates. The use of the term *distinct* in the previous sentence emphasizes the fact that repetitions of the same result are of no import. These issues are treated further in Appendix B.

From an operational viewpoint, however, the closed nature of the axiom set has little practical consequence. The human mind is limited in its capacity to handle a multi-faceted problem concurrently, and in its speed of operation. Hence, a computerized system would be of assistance even if its capabilities were no greater than that of a human. The computer-based system, however, does have the advantages of speed, patience, and thoroughness.

To the extent that symbolic logic effectively models human reasoning processes, the constraints on an automated system are no greater than those on a person.

According to this perspective, an automated system can be at least as capable as a person. The next section discusses some design criteria for software systems.

## Practical Implications

From an engineering perspective, the utility of a general model of design relates directly to its practical consequences. We now explore the implications of the transformation model of design and its formalization as the basis for design automation.

### Desiderata for Software

An ideal software methodology should be characterized by a number of factors. These attributes are as follows:

- *Correctness.* The software should be logically valid; it should be *sound* in that *only* true statements are derived, and complete (or expressible) in that *all* true results can be derived.

- *Validity*. A conclusion must be correct not only in a strictly logical sense, but also in the context of the application. For example, two manipulators of a robot may have intersecting workspaces; but the appendages should not be required to occupy the same space at the same time. Hence, we may speak of soundness and completeness in terms of domain validity, just as for logical correctness above.
- *Modularity*. The methodology should incorporate a modular architecture to allow for the containment of complexity. This promotes ease of user comprehension as well as structural modifications.
- *Flexibility*. A software methodology should be flexible or convenient to use in the following senses:
  *Power*. The package should allow for economy of expression for the implementation of high-level concepts.
  *Range*. Low-level, detailed specifications should be describable when appropriate.
  *Extensibility*. The methodology should readily allow for the incorporation of new concepts. For example, the concept of an automobile "engine" should be definable from individual components, just as the notion of a "read" macrocommand can be specified as a sequence of elementary input/output computer operations.
- *Friendliness*. This characteristic may be subdivided into two categories.
  *Explanation*. Explanation facilities should be available to justify the reasoning processes in order to convince skeptical end-users and to facilitate modifications by the system developers.
  *Interfacing*. The system should provide for high level interactions in concepts familiar to the end-user. Examples relate to graphic icons and natural language interpretors.
- *Efficiency*. The results should be delivered in timely fashion, and without the need for excessive resource requirements.

A reasoning strategy based on mathematical logic will be correct (sound and complete) to the extent that predicate calculus is correct. The friendliness of a package may be ensured by using techniques such as explanation modules and natural language—technologies that are widely available.

The validity of the reasoning process depends, by definition, on the domain of application. This attribute is discussed further in the following subsection, in conjunction with software tractability issues relating to modularity and efficiency.

## Validity and Tractability

The formal model of the design process is a proper tool for studying the fundamental nature and limits to synthetic activities. However, if the general paradigm is to be implemented in software, then we must consider as well issues relating to computational complexity: it is imperative to consider not only what is possible but also what is practical.

In general, a forward-chaining approach from the initial axioms to all relevant theorems is inadequate due to the combinational explosion in derivable theorems. Hence the reasoning process must be guided by metalevel knowledge. Such knowledge may be of two types: general procedures and domain-specific knowledge.

*General procedures* are domain-independent rules to steer the reasoning process. For instance, forward chaining is preferable to backward chaining if the initial knowledge base is less numerous than the potential goal states and vice versa. As a second example, a sequence of reasoning steps should be pursued further if the results are increasingly specific and seem likely to converge quickly to a particular implementation.

*Domain-specific knowledge* refers to heuristics relating to different areas of application. The organizational structure might be hierarchical, as depicted in Figure A.4. Each node in the figure may be decomposed further at will, as exemplified by the partitioning of knowledge for the manufacturing environment.[16]

Current research on generic design methodologies relates to systematic frameworks and decision rules to address both these issues. One such method, called the axiomatic approach, was introduced previously and is discussed further in Appendix C.

## Future Directions

Future work toward a science of design may be categorized into theoretical and practical concerns. The development of a *theory* refers to the systematic, formal representation of design activity. This class of objectives may in turn be grouped into representational techniques and additional principles.

The modeling of transitions in design as quintuples raises a number of concerns relating to the completeness and uniqueness of the representation. For instance, the set of initial states should be small enough to be manageable, yet rich enough to permit proper reasoning. Moreover, the representation of the final state should be general enough to encompass diverse design situations. Another issue relates to the fact that a design problem usually admits more than one solution. This situation underscores the need for a procedure to generate an appropriate subset of solutions and select the best among them.

In addition to the overall capabilities of the formal models, we need to better understand the common elements in the design process across diverse realms. Further theorems are needed to better define the structure, organization, control, and implementation of design techniques. The long-term goal of this line of work is to develop a complete set of axioms and theorems that captures strategies used by human experts in various domains. This knowledge would augment basic domain knowledge as well as the stock of transformation operators.

*Practical* concerns relate to the implementation issues. The framework and approaches discussed previously may serve as the basis for a generic system for design. A proposed architecture for such a system is shown in Figure A.7.[17] Initially, the system may serve as a consultive advisor to a human designer. With increasing capability, the package may proceed further and propose solutions autonomously in various domains of application.

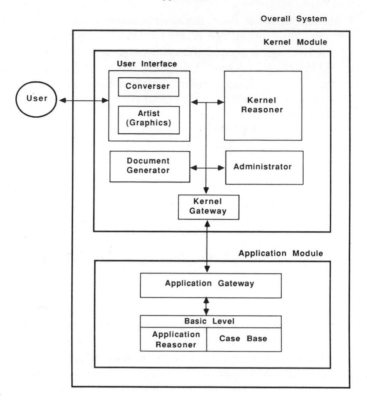

**Figure A.7** Software architecture for a knowledge-based system for design.

The two directions presented above are mutually supportive. Progress in the theoretical arena will support implementation decisions, just as practical considerations may be integrated and resolved through abstract representations.

## Summary

The design process may be modeled as a sequence of transitions, from an initial knowledge base to the final state which includes the desired result. In terms of mathematical logic, this process may be described as an *initial set* of axioms that are transformed into a *final set* of statements through a *transformation* process. The transformation process may be viewed as the creation or elimination of theorems through a set of metalevel *operators*. The consequences of formalization relate to the clarification of the nature of the design process, and implications for design automation. The utility of the formal model was illustrated through its interpretation in the context of discovery in design, followed by a discussion of the limits to such discovery. By incorporating this model in a knowledge-based system, it may also serve as a basis for an automated designer.

# APPENDIX B

# Predeterminism in
# Reasoning and Creativity

*Mankind and the animals with analogous abilities are distinguished by their capacity for the introduction of novelty. This requires a conceptual power which can imagine, and a practical power which can effect.*[1]

Alfred North Whitehead

The generation of new designs—whether in the form of a physical product, organizational structure, or mathematical proof—is characterized by differing levels of surprise. Sometimes we are astonished at the novelty of the result, and at other times unimpressed.

We are astonished when the result is so new that we do not readily understand the process by which it was generated. This forward procedure may take us by surprise, but the backward procedure of rationalization usually does not. In other words, we are reassured once we understand the forward reasoning steps and can rationalize why the result is consistent with—and implied by—the initial conditions and functional requirements.

When we do not comprehend the "leaps of intuition" that can occur in generating novel designs, we call the process creative. But to the extent that such creative acts can be understood in retrospect, they are no more logical or illogical than other forms of rational thinking.

The fruits of reasoning and creativity are syntactically "closed" in the sense that whatever results we derive are consequences of both our knowledge of previous facts and accepted rules of inference. On the other hand, these results are semantically "open" in the sense that we come to recognize new facts and relationships that were not previously apparent.

A problem from high school algebra is the following:

Yvonne drives for 3 hours at an average speed of 50 kilometers per hour. She takes the same route on her return trip and completes it in 2 hours. If the time spent at the destination is negligible, what is Yvonne's average speed over the entire trip?

The facts of the case do not include the right answer. By the rules of algebra, however, one can reason that Yvonne drove 300 kilometers in 5 hours, from which her average speed is 60 kilometers per hour. The open/closed nature of rational

195

thought is applicable not only to specific domains such as algebra, but to the entire spectrum of human cerebration.

The process of reasoning or deliberate thinking can be modeled using the formalisms of predicate calculus. This appendix shows that although new results can be generated through "creative" processes, the results are closed in the sense that no new predicate can be generated. The presentation is based on earlier work by the author.[2]

# Symbolic Logic

The next two sections review the two complementary perspectives in symbolic logic, relating to the syntactic and semantic interpretations.[3] This is followed by a discussion of the equivalence of the two approaches, and a formal discussion of the nature of predetermination in reasoning.

People use logic as a matter of course in everyday life. For example, a person who knows that a falling barometer level indicates the approach of foul weather will draw the right conclusion from observing the slide in atmospheric pressure.

Unfortunately, we humans are prone to error when it comes to complex problems in reasoning. For example, careful studies of legal documents—which are intended to be paragons of logical precision—often result in the exposure of internal contradictions.

Whether by design or by inculcation, we are often careless in stating assumptions explicitly and sloppy in performing deductions. Although our long-term memory capacity seems to be infinite in practical terms, our short-term working memory can handle only about seven items concurrently. Further, although our minds are capable of occasional leaps of intuition (which may or may not turn out to be valid), they are remarkably slow in making even routine deductions.

Mathematical logic, also called symbolic logic, was developed to address the problems of faulty reasoning and ambiguity in everyday language. The basic premise is that correct procedures for reasoning or deductive thinking can be formalized and used profitably, independently of the specific domain of discourse.

# Semantic Viewpoint in Logic

This section first introduces the primitive concepts of propositional calculus, then presents its generalization known as predicate calculus.

## *Propositional Logic*

*Propositional logic* or *calculus* is a subset of mathematical logic that deals with simple declarative expressions or sentences. These sentences are called *propositions*, as exemplified by the statement "Time flies like an arrow" or "Students like to party."

Each proposition is considered to be *true* or *false*; these *truth values* may be abbreviated simply to *t* and *f*. Individual propositions may be combined into more elaborate propositions by using symbols called *functors* or *connectors*, as defined below.

The *and* functor, called *conjunction* and represented by the symbol &, combines two propositions *P* and *Q* in the following way: the compound proposition (*P* & *Q*) is regarded to be true if both *P* and *Q* are true, and is false otherwise. To illustrate, let *S* denote the proposition "The Sun is a medium-sized star"; *E* the proposition "The Earth is flat"; and *M* the proposition "Mars has a thin atmosphere." To the best of our knowledge, *S* is true, *E* false, and *M* true. The compound predicate (*S* & *M*) is true since each of *S* and *M* is true, while (*S* & *E*) is false because at least one of them (*E*) is false.

The characteristics of the *and* functor are also defined by the first three columns of Table B.1. For example, the second row specifies that if proposition *P* is true and *Q* is false, then (*P* & *Q*) is false.

The *or* functor, called *disjunction* and symbolized as $\vee$, combines two propositions *P* and *Q* into a new proposition in the following way: (*P* $\vee$ *Q*) is true if at least one of *P* or *Q* is true, and false otherwise. To illustrate by using the previously defined propositions, *S* $\vee$ *E* is true since at least one proposition (*S*) is true. The characteristics of the *or* functor are depicted by the first, second, and fourth columns of Table B.1.

The *not* functor, called *negation* and symbolized as $\sim$, reverses the truth value of its referent proposition. In other words, $\sim P$ is true when *P* is false, and is false when *P* is true. The first and fifth columns of Table B.1 define this relationship.

The *if-then* functor, called *implication* and symbolized by a right arrow $\rightarrow$ has the following property: (*P* $\rightarrow$ *Q*) is false when *P* is true and *Q* is false, and is true in all other cases. This characteristic is reflected in the first, second, and sixth columns of Table B.1.

To illustrate, let *M* denote the proposition "Cinderella will marry the prince" and *H* the proposition "Cinderella will live happily ever after." Suppose we are told that (*M* $\rightarrow$ *H*) is true: Cinderella will live happily if she marries the prince. Then if we observe Cindy marrying the prince, we know that she will find bliss. But what happens if we observe the dastardly prince marrying someone else? We cannot determine Cindy's disposition in her old age. In other words, (*M* $\rightarrow$ *H*) does not preclude the possibility that she lives blissfully in solitude; or that she marries Joe, the boy next door, and lives happily; or perhaps she chooses Jim and finds misery. This is the rationale for defining the truth values of the implication functor in a way which may seem at first sight to be counterintuitive.

*Table B.1* Logical functors and truth values

| P | Q | P & Q | P $\vee$ Q | $\sim$ P | P $\rightarrow$ Q |
|---|---|---|---|---|---|
| *f* | *f* | *f* | *f* | *t* | *t* |
| *f* | *t* | *f* | *t* | *t* | *t* |
| *t* | *f* | *f* | *t* | *f* | *f* |
| *t* | *t* | *t* | *t* | *f* | *t* |

The *if-and-only-if* functor, also called *equivalence* or simply *iff* and symbolized as ↔, is defined in the following way: $(P ↔ Q)$ is true if both $(P → Q)$ and $(Q → P)$ are true; and is false otherwise. This is tantamount to saying that $(P ↔ Q)$ is true precisely when $P$ and $Q$ are both true or both false.

## Predicate Logic

While propositional logic has many uses, it deals in categorical statements about sets of objects and does not allow for variation within those sets. For example, propositional calculus allows for statements such as "Students hate opera" or even "Julie hates opera." However, it precludes statements such as "If $x$ is a senior, then $x$ will attend the Senior Prom" for some arbitrary student $x$.

*Predicate logic* or *calculus* is a generalization of propositional logic that allows for variables. A *predicate* is a mapping from a set of objects to the truth values $t$ and $f$.

To illustrate, let $S(x)$ denote the predicate "Is a senior"; then $S(x)$ is true if $x$ is a senior and false otherwise. Similarly, let $Pr(x)$ denote the predicate "will attend the Senior Prom." Then the statement "If $x$ is a senior, $x$ will attend the Senior Prom" may be written as $S(x) → Pr(x)$.

Often we would like to be more specific and stipulate whether a predicate is valid for some or all objects of a given set. This is achieved through the use of quantifiers.

The *existential* quantifier, denoted ∃ and read as "there exists," is defined as follows: $∃x\ P(x)$ is true if there is at least one object $x$ for which predicate $P$ is true, and is false otherwise. Using the previous example, $∃x\ Pr(x)$ is true if there is some person going to the prom, and is false if there is none.

The *universal* quantifier, denoted as ∀ and read "for all," is defined in the following way: $∀x\ P(x)$ is true if $P(x)$ is true for all $x$, and is false otherwise. For example, $∀x\ Pr(x)$ is true if everyone is going to the prom, and is false if at least one person abstains.

These structures may be used to formalize principles or decision rules such as those at the end of each chapter in Part II of this book. A decision rule relating to coordination specifies when a system should be decentralized rather than regimented from a single location. The statement

$$∀s\ ∀e\ \ \{\text{systemInEnv}(s,e)\ \&$$
$$[\text{coordinativeCapability}(s) < \text{diversity}(e) \lor$$
$$\text{communicationDelays}(s) > \text{tempo}(e)]$$
$$→ \text{decentralize}(s)\}$$

may be used to encode the notion that, for any system $s$ in environment $e$, it should be decentralized if its coordinative capacity is less than the diversity of $e$, or its internal communication delays exceed the tempo of $e$.

## Semantic Implication

In the semantic approach to logic, each statement—also called an *expression, formula,* or *well-formed formula*—is viewed to be true or false. Since variables are not allowed in propositional calculus, an interpretation is simply an assignment of truth values to atomic formulas. These truth values may then be used to determine the truth values of more involved formulas or statements.

The definition of interpretation must be modified for first-order predicate calculus, which does allow for variables. This may be done in terms of a domain and an assignment of truth values to the symbols in a formula, whether they represent constants, functions, or predicates. The notion of interpretation is made more precise in the following definition, which consists of three parts:

> **Definition.** An *interpretation* of a formula consists of a nonempty domain $D$ and an assignment of "values" to each constant or function symbol in the formula in the following way:
>
> 1. Each constant symbol is assigned an element in $D$.
>
> 2. Each $n$-place function symbol is assigned a mapping from $D^n$ to $D$, where $D^n$ is defined as the $n$th order Cartesian product of $D$. In other words,
>
> $$D^n = \{\langle x_1, \ldots, x_n \rangle \mid x_i \in D \text{ for } i = 1, \ldots, n\}$$
>
> 3. Each predicate symbol is assigned a mapping from $D^n$ to $\{t, f\}$, where $t$ represents *true* and $f$ represents *false*.

Thus a predicate symbol is a special kind of function whose co-domain consists of exactly two objects, $t$ and $f$.

An open formula is one that contains free variables. Hence such a formula may be true or false depending on the specific assignments of constants to variables. To obtain categorical or absolute results rather than conditional ones, we work with closed formulas in the semantic approach to logic. In this way a formula is either true or false in a given interpretation, regardless of the values assigned to its variables.

**Definition.**
1. An interpretation of a set $W = \{W_i\}$ of formulas is called a *model* of $W$ iff every $W_i$ is true in that interpretation.
2. A formula $W$ is a *logical consequence* of a set of formulas $W$ iff $W$ is true in all models of $W$. In this case, we write

$$W \vDash W$$

3. A formula $W$ is *satisfiable* if it has a model; otherwise it is *unsatisfiable*.
4. A formula $W$ is *valid* if it is true in all possible interpretations; otherwise it is invalid.

Thus if a formula is true in some interpretations but not all, it is said to be satisfiable but invalid.

## Syntactic Viewpoint in Logic

Certain formulas are necessarily valid due to the meaning of the logical symbols, as exemplified by the expression $x \leftrightarrow x$. Formulas such as these are called *logical axioms*. Other axioms which are valid by hypothesis are called *proper* or *nonlogical axioms*.

Predicate calculus consists of logical axioms and two production rules. These rules of inference are

1. *Modus Ponens:* From $P$ and $(P \rightarrow Q)$, infer $Q$.
2. *Generalization:* From $P$, infer $(\forall x)P$.

Inference rules may be used to deduce theorems from axioms. A first-order theory $T$ is a formal system consisting of the logical and nonlogical axioms as well as the production rules.

When a theorem $W$ is deducible from the axioms of a theory $T$, we write

$$\vdash_T W$$

or simply

$$\vdash W$$

when the theory $T$ is apparent from the context or is inconsequential. A formula $W$ is a consequence of a set of formulas $W$ if $W$ is derivable from the axioms of theory $T$ and formulas in $W$. In this case we write

$$W \vdash_T W$$

A model of $T$ is by definition an interpretation in which all axioms of $T$ are true. A nonobvious property of a set of axioms $T$ is that a theorem derivable from $T$ will be true in every model of that set.

## Soundness versus Completeness

The semantic approach focuses on the validity of a formula, while the syntactic approach deals with the construction of proofs of formulas. These viewpoints, however, differ only superficially, as they can be shown to be equivalent. This equivalence is known as the completeness and soundness results.

$$W \vDash W \quad \underset{\text{Soundness}}{\overset{\text{Completeness}}{\rightleftarrows}} \quad W \vdash W$$

*Completeness* refers to the fact that if a formula $W$ is true in all models of the set of axioms $W$, then $W$ can be proved from $W$ viewed as a set of hypotheses. On the other hand, *soundness* pertains to the fact that if the formula $W$ is derivable from a set of axioms $W$, then $W$ is true in all models of $W$.

## Predetermination in Thinking

Logical systems are constrained in the degree to which new results may be derived. The following metatheorem states that no new predicate can be derived from a set of axioms that does not contain such a predicate.

**Metatheorem** (Nondeducibility). Let $A$ be a set of axioms, none of which includes the predicate $P$. Then no theorem which includes $P$ can be derived from $A$.

The proof is based on the idea that, according to the syntactic viewpoint discussed in an earlier section of this appendix, neither of the two methods of generating new results—*modus ponens* and generalization—allows for the creation of new predicates. The lack of new predicates implies that no new attributes or relationships among objects can be created, for those that were not previously implied by the initial knowledge base. Further, the equivalence of the syntactic and semantic viewpoints, as explained in the previous section, implies that the semantic approach will not contain any greater power of expression. (The details of the proof are available elsewhere.[4])

The closed nature of deliberation refers to meaningful deductions generated by a reasoning process. This conclusion is valid in the context of the problem domain.

If relevance to the problem domain were not at issue, we could readily devise trivial procedures for an unbounded number of deductions. These "results" would follow directly from the definitions of logical operators.

For example, consider a reasoning system to control an autonomous rover on Mars. Let $M$ denote the proposition "Mars has a rocky terrain" and $P$, the sentence "One minus one is zero." If $M$ and $P$ are considered to be true, then so are all the following:

$$\sim\sim M$$
$$M \vee P$$
$$M \,\&\, P$$
$$M \vee P$$
$$M \vee P \vee \sim P \vee \sim M$$

as well as an infinitude of others. However, these latter sentences contribute nothing to the rover's ability to reason about its environment.

The preceding discussion would suggest that creativity is not an entirely mysterious phenomenon. In fact, the terms *intuition* or *creativity* refer to the process in which the human mind arrives at a tentative conclusion while side-stepping reasoning steps. At times the flights of intuition can be justified in retrospect through logical deductions; at other times the flights have little or no grounding in the facts of the case.

The closed nature of a deductive system has theoretical implications, but little in the way of practical constraints. This viewpoint was discussed in detail in Appendix A.

## Openness and Domain-dependence

An intelligent agent should be able to learn from its experiences rather than remain forever captive to its initial set of facts and inference procedures. In this way, the agent may break free of its original limitations, improve in performance over time, and perhaps even generate new results that might be labeled creative.

A domain-dependent learning system is one that can fuse properties and relationships to objects in novel ways. The objects in question may already exist in the object base $W$; or they may be newly introduced as the result of the agent's encounter with a wholly new phenomenon. In a similar way, the properties and relationships may be entirely novel, or may result from the juxtaposition of existing elements.

Consider an automated system for configuring a factory layout, given information on departments, flows, space requirements, and others. A comparison of acceptable versus unacceptable designs may lead to a new heuristic based on Wall 3: "Try to locate offices near Wall 3." This situation might occur, for example, if the shipping and receiving departments are located next to the roadway on the east side of the plant; the transfer lines begin and end on the east; offices cannot be near conveyors due to noise; the west side of the factory has been labeled Wall 3. Further experience in plant design might lead to a revision of the above rule through the notion of a "large" factory ("Offices need not be near Wall 3 if the factory is large") or a "quiet" conveyor system ("Offices need not be near Wall 3 if the conveyor system is quiet").

The ability to manipulate a knowledge base requires capability beyond the first-order predicate logic presented earlier. The power to reason about first-order statements takes us into second-order predicate logic, a facility that is readily available in certain computer languages such as Prolog.

Learning systems based on connectionist models incorporate their newfound knowledge implicitly, in the form of weights in nodes and arcs. In contrast, other systems represent their knowledge explicitly, such as rule-based programs written in Prolog. A rule-based system that learns from its domain experience requires the ability to modify its knowledge base, and thereby possesses the expressive power of second-order logic or higher.

## Summary

This appendix presented an introduction to the syntactic versus semantic perspectives of mathematical logic. The material served as background for the Nondeducibility Metatheorem: a deductive system is closed in the sense that no new predicates can be generated by a logical system operating on a set of axioms or facts.

The discussion has explored the nature of reasoning and learning in correspondence to the knowledge base and the degree of domain independence. We may conclude the following:

- The knowledge base of an intelligent system may be formalized in terms of a simple algebraic model.
- Inference procedures based on first-order predicate logic lead to domain-independent results that are not influenced by the experiences of the agent. These methods imply the *closed* nature of reasoning and creativity.

- A truly adaptive system must be able to augment its knowledge base as a result of its observations of reality.
- Procedures for generating new knowledge which transcends straightforward logical implications, must be based on domain-dependent facts. In other words, the open nature of a reasoning system is directly correlated with the level of domain dependence.

Future work should explore both theoretical and practical concerns. On the theoretical side, we need to investigate in greater detail the nature of openness and the differing levels of domain dependence. Under what conditions is it permissible to invoke new hypotheses?[6] How can this best be done? How can the various hypotheses, some of which may seem to conflict, be integrated into a rational strategy for action?

In a practical vein, research is required on computational methods for implementing creative, novel thinking, and on their implications for attaining effective results. A closely related issue is that of complexity—the relative efficiency in terms of time and memory requirements. The ultimate objective in practical terms is clear: to develop useful systems such as fully autonomous robots and automated designers capable of generating novel products.

# APPENDIX C

# Axiomatic Approach to Design

*There are and can be only two ways of searching into and discovering truth. The one flies from the senses and the particulars to the most general axioms, and from these principles, the truth of which it takes for settled and immoveable, proceeds to judgement and to the discovery of middle axioms. And this way is now in fashion. The other derives axioms from the senses and particulars, rising by a gradual and unbroken ascent, so that it arrives at the most general axioms last of all. This is the true way.*[1]

Francis Bacon

The role of engineering is to satisfy human needs by creating solutions to problems. These solutions may take the form of products, processes, or systems that incorporate the prevailing body of scientific and technological knowledge. Such activities involve the synthesis of a solution, followed by analysis for evaluating and optimizing the product of the synthesis phase. Since these phases often proceed concurrently, improvements in analytical and synthetic techniques have a synergistic impact on the quality of the design activity.

Due to extensive research in the area, numerous principles and methodologies exist for the analysis of physical systems. In contrast, few tools are available for the synthesis phase, as it is poorly understood.

A design problem begins with the perception of a human need. This need is translated into a set of functional requirements that must be fulfilled by some appropriate collection of design parameters. The quest for the proper set of parameters is an open-ended selection process that may be interpreted as a forward-chaining search procedure. This interpretation holds when the activity is viewed as a search that starts from a set of functional requirements and moves forward, generating alternative sets of design parameters until a satisfactory one is found. The sheer volume of the search space precludes the use of brute force to solve problems of realistic complexity. The search tree must be pruned by infusing experience and domain knowledge into the system.

To this end, a general set of guidelines has been proposed in the form of the Design Axioms. These Axioms are generic decision rules for the design of products and processes. They have been applied with remarkable success in applications ranging from machine design to polymer processing. For a given application area, the Axioms may be augmented by guidelines that are even more specific. For example, a consultive system for designing robot components may refer to a knowledge base incorporating more specific heuristics as well as data. The Design Axioms have been formalized using mathematical logic, then translated into the logical language of Prolog to serve as the basis for a consultive expert system.[2]

The Axioms have been advanced as a foundation for the creative phase by the engineering scientist Nam Suh and his colleagues.[3] The Design Axioms consist of two general postulates, stated originally in the following form:

> **Axiom 1** (Independence).   Maintain the independence of functional requirements.
>
> **Axiom 2** (Information).   Minimize information content.

Implicit in the statement of the second Axiom is its dependence on the first; that is, information is to be minimized subject to the fulfillment of functional requirements.

To illustrate the use of Axiom 1, consider a mobile robot that is to serve as a camping and hiking companion. Its functional requirements might be the following:

1. Traverse rough terrain having obstacles up to 50 cm.
2. Monitor distance traveled up to 1000 km with an accuracy of 1 m.
3. Carry a payload of 200 kg.

The fulfillment of each of these functional requirements should be addressed by different structural elements in the robot. For example, the joint members that provide for flexibility in the mobility system should not be the primary structures that support the payload. In a similar way, the angular displacements of the legs or other mechanisms whose prime responsibility is to propel the robot forward should not serve as the basis for the odometer, for which accuracy is the key.

The decomposition of a corporation into divisions is a further example of the Independence Axiom. Suppose that the functional requirements of the company are to produce radios, shoes, and cookies. These goods are best handled by relatively autonomous divisions within the corporation so that the production requirement on cookies, for example, is unaffected by the factory capacity for radios or the marketing activites for shoes.

The Information Axiom calls for a minimization of information requirements. Returning to the robotic example, if the odometer requires an accuracy of $\pm 1$m in 1000 km, then the designer should not stipulate tolerances so tight that the range finder is accurate to $\pm 1$ cm. To do so would increase system complexity, compromise reliability, raise production cost, or perhaps produce some other side-effect.

The Design Axioms are generalized decision principles for use in the synthesis phase. They encapsulate empirical phenomena in all aspects of synthesis such as those encountered in product design and manufacturing. In this sense they are equivalent to propositions that are called laws in the physical sciences, such as the laws of gravitation or thermodynamics. The significance and implications of the Design Axioms are discussed in greater detail in the literature.[4]

If the axioms and their corollaries are to be stated precisely and their interrelationships determined rigorously, then they must be expressed in mathematical form.

The next section addresses the way in which symbolic logic may be used for this purpose, while the subsequent sections discuss the consequences of formalization for encoding the axioms into software. This appendix is based on a number of previous publications.[5]

## Design Axioms in Predicate Logic

A design that satisfies its referent functional requirements and constraints is said to be feasible. We define the predicate *feas*(•) in the following way: for a design *x*, *feas(x)* is true if *x* is feasible, and is false otherwise.

A design is said to be *coupled* if the functional requirements cannot be satisfied independently; in other words, the fulfillment of one functional requirement interferes with that of another. On the other hand, a design whose functions can be met independently is called *uncoupled*.

Many physical phenomena exhibit inherent coupling when they are subject to natural or conventional methods of treatment. For example, the functional requirements of an extruded polymer sheet may pertain to thickness and density. Unfortunately, these latter parameters are coupled in most conventional processing techniques, thereby hampering or even preventing the fulfillment of the original functional requirements. The first design principle then calls for a search for an uncoupled design. Often the uncoupled design may be simple in itself but is not readily apparent. Further discussion as well as actual examples are available in publications by Suh.[6]

Let *coup(x)* be the predicate asserting that design *x* is coupled, while *unc(x)* is defined as $\sim coup(x)$. In addition, *acc(x)* means that design *x* is acceptable, while *super(x,y)* holds if and only if design *x* is superior to design *y*. The first axiom may be stated in symbolic logic in the following way:

**Axiom 1.** A feasible design which is uncoupled, is acceptable:

$$\forall x \; \{feas(x) \; \& \; unc(x) \rightarrow acc(x)\}.$$

The universal quantifier is required to assert the generality of this statement. The expression in braces states that if some object is feasible and uncoupled, then it is acceptable. The universal quantification stipulates, moreover, that this statement holds for *all* such objects rather than merely for *some* particular object.

The second axiom relates to the informational (as opposed to functional) complexity of designs. Let *ifm*(•) denote a measure of information content defined on the set of feasible designs. Then $ifm(x) < ifm(y)$ would imply that the information content of design *x* is less than that of *y*. For convenience we sometimes use such infix notation to stand for the associated predicate. For example, we write

$$ifm(x) < ifm(y)$$

to denote the predicate

$$lessThan(ifm(x), \; ifm(y))$$

which is true if and only if *ifm(x)* is less than *ifm(y)*.

The second Axiom is as follows:

**Axiom 2.** Of two acceptable designs, the one with less information is superior.

$$\forall\, x\ \forall\, y\ \{acc(x)\ \&\ acc(y)\ \&\ ifm(x) < ifm(y)$$
$$\rightarrow\ super(x,\ y)\}.$$

A number of propositions were included in the original list of decision rules proposed in the axiomatic approach to design.[7] Subsequent research indicated that the Independence and Information Axioms are the key concepts, and that the other decision rules follow as corollaries.

## Direct Consequence

The following proposition is an immediate consequence of the Independence Axiom.

**Proposition** (Decoupling). Decouple functions that are coupled.

The Decoupling Corollary may be stated more precisely in the following way. Let $u$ be a feasible design which is uncoupled; then we have the following facts:

1. *feas(u).*
2. *unc(u).*

If we instantiate $x$ to $u$ in Axiom 1, the result is

3. *feas(u) & unc(u) $\rightarrow$ acc(u).*

These three items yield the conclusion

4. *acc(u).*

This Proposition is related solely to Axiom 1. The only way for a theorem to be provable from a single axiom is for it to have the same formal structure as the axiom, as is the case here. In other words, the Decoupling Proposition is a corollary or alternative statement of Axiom 1 in an informal sense, but is strictly a restatement of the Axiom in a formal sense. Hence it would be more appropriate to call this proposition an alternative informal statement of the Independence Axiom rather than a corollary.

The Decoupling Proposition can be applied to a spectrum of domains ranging from the design of products to entire organizations. For example, a decision rule from the realm of factory planning is:

**Proposition** (Departmental Configuration).
Decouple the functional roles of departments.

## Indirect Consequence

Most other corollaries are not obtainable solely from the axioms. With some plausible assumptions, however, they may be justified as indirect consequences.

The first corollary relates to the design of process plans. It depends on the reasonable assumption that information can be decreased by grouping similar operations, thereby minimizing backtracking. This postulate may be written, using obvious choices for predicates and functions, as

1. *clustering(u)* > *clustering(v)*
   → *ifm(u)* < *ifm(v)*.

The associated corollary may be stated as:

> **Corollary** (Clustering). Cluster similar production operations.
>
> $\forall x \, \forall y \, \{acc(x) \,\&\, acc(y) \,\&\, clustering(x) > clustering(y)$
>     → *super(x,y)*}.

By instantiating *x* to *u* and *y* to *v*, the antecedent is composed of the following facts:

2. *acc(u)*.
3. *acc(v)*.
4. *clustering(u)* > *clustering(v)*.

From items 1 and 4, we obtain

5. *ifm(u)* < *ifm(v)*.

We instantiate *x* to *u* and *y* to *v* in Axiom 2, giving

6. *acc(u)* & *acc(v)* & *ifm(u)* < *ifm(v)*
   → *super(u,v)*.

Items 2, 3, and 5 in conjunction with item 6 imply

7. *super(u,v)*.

which is the consequent of the Clustering Corollary. Since *u* and *v* are arbitrary process designs (which satisfy the antecedent), the result is valid for all *x* and *y*. Hence the Corollary follows.

The information required for a task is based on the probablility of success. The requisite information decreases as the likelihood of attainment increases (see Appendix E). We may state this relationship in the following formal sentence:

$$\forall x \, \forall y \, \{prob(x) > prob(y) \rightarrow ifm(x) < ifm(y)\}.$$

In other words, if a design *x* has greater likelihood of success than that of *y*, then its information requirement is less. This is a domain-independent rule that is based purely on the definition of information.

In a turbulent economic environment where the rate of innovation is high, product lifespans short, and lot sizes small, the long-term viability of a manufacturing plant increases with its flexibility. Hence we may write:

$$\forall x \, \forall y \, \{flex(x) > flex(y) \rightarrow prob(x) > prob(y)\}.$$

That is, when the flexibility of a plant *x* exceeds that of *y*, then its probability of success is higher. This sentence is a domain-specific rule pertaining to the field of manufacturing plants. In particular, this rule holds for industrial plants operating in turbulent markets subject to rapid product innovation.

The preceding two rules may be used in conjunction with the design axioms to conclude that a flexible plant is better than an inflexible one due to its reduced need for prior information concerning the environment.

A number of decision rules for factory design can be derived from the Information Axiom and the assumption that information needs expand with physical distance. The supporting assumption may be stated as follows:

> **Proposition** (Physical Distance). Given 2 pairs of interacting stations that incorporate equivalent functions, the pair with less separation requires less information.

By a reasoning process similar to that of the Clustering Corollary, decision rules such as the following may be derived:[8]

> **Corollary** (Receiving/Starting). Locate the receiving area close to the start of the production process.

> **Corollary** (Assembly/Packaging). Locate the final assembly area near the packaging station.

The introduction of other reasonable assumptions leads to consequences such as the following:

> **Corollary** (Walking). Minimize walking distances required of operators.

> **Corollary** (Length). Minimize the maximum length of each supervisor's area.

As suggested earlier, the general strategy for formalizing and validating guidelines for factory planning may be applied to other realms of design. More specifically, the following corollaries have been shown to be derivable from the axioms and some innocuous assumptions.[9]

> **Corollary** (Processing). Minimize processing information.

> **Corollary** (Conservation). Conserve materials and energy.

> **Corollary** (Weakness). If weaknesses cannot be avoided, separate parts.

> **Corollary** (Integration). Integrate components if functional independence is not impaired.

> **Corollary** (Part Count). Part count is not a measure of productivity.

> **Corollary** (Standard Parts). Use standard or interchangeable parts and processes whenever possible.

## Design Axioms in Prolog

Prolog (Programming in logic) is a very high-level programming language based on mathematical logic.[10] This section demonstrates how the formalization of the Design Axioms in symbolic logic readily allows for their translation into software through Prolog.

## Axioms in Prolog

Axiom 1 may be written in Prolog as

$$\textbf{acc(X) :-}$$
$$\textbf{feas(X),}$$
$$\textbf{unc(X).}$$

This says that design **X** is an acceptable one if it is uncoupled.

Let **iff(X, Ifm)** be the predicate that maps a design **X** into its overall information measure **Ifm.** Then Axiom 2 may be given as

$$\textbf{super(X,Y) :-}$$
$$\textbf{acc(X),}$$
$$\textbf{acc(Y),}$$
$$\textbf{iff(X, Ifmx),}$$
$$\textbf{iff(Y, Ifmy),}$$
$$\textbf{Ifmx} < \textbf{Ifmy.}$$

The verbal translation is that design **X** is superior to design **Y** if both are acceptable (by Axiom 1) and design **X** has less information content than **Y**.

## Data Structures for Axioms 1 and 2

A generalized data structure to encode design data takes the form **D(X, L)** where **D** represents the name of a given attribute of design **X** and **L** the corresponding list of specifications. For the case where **L** is an empty set, we may write **D(X)** rather than **D(X, [ ] )**. The specializations of this general structure are discussed below.

*Information Axiom.* Consider the geometric information relating to a rectangular block. Let **U, V,** and **W** represent lengths along three spatial axes, with respective tolerances **DU, DV,** and **DW.** Then these specifications might be encoded in the form

$$\textbf{geomd(X, [U,DU,V,DV,W,DW] ).}$$

We need a way to calculate the information value of this specification. Let **geomf(X, Geomm)** be the function which maps a design **X** into its geometric information measure **Geomm.** A reasonable procedure for **geomf** is given by

$$\textbf{geomf(X, Geomm) :-}$$
$$\textbf{geomd(X, [U,DU,V,DV,W,DW] ),}$$
$$\textbf{Geomm is (log (U/DU) + log (V/DV) + log (W/DW)).}$$

where **log** denotes the logarithm to base 2.

The data structure **D(X,L)** is a sufficient but not necessarily desirable construct for encoding design specifications. When an object has numerous attributes, for example, then a frame representation may be more efficient in terms of memory and more aligned with users' conceptions.

The *frame* is a data structure of the form:

$$F :: L$$

where the quantity $F$ is a label for the frame, and $L$ is a *list* of data items. More specifically, $L$ is a list of the form

$$[S_1 : V_1,$$
$$S_2 : V_2,$$
$$\vdots$$
$$S_n : V_n]$$

where each term $S_i$ denotes a *slot* or *attribute* of $F$ and $V_i$ the corresponding *value*.

To illustrate, a consultive expert system for designing robotic manipulators might contain a frame such as the following:

handler :: [ unit-type    :    manipulator,

max-width    :    20 cm,

precision    :    25 micrometers,

max-load    :    2 kg ]

This frame would stipulate that the handler is a manipulator unit with a maximum extension of 20 cm, a precision of 25 micrometers, and maximum load of 2 kg. Structures such as frames can be readily implemented in Prolog using facilities for defining operators.

In general, a software package will incorporate more than one type of knowledge structure, to facilitate human comprehension or processing efficiency.

*Independence Axiom.* We would ideally like to have a computer procedure that takes a design and determines whether or not it is coupled. Unfortunately, such a procedure cannot be readily implemented at the current state of the art, since the field of artificial intelligence is not sufficiently developed to permit the automatic deduction of such knowledge from design data for arbitrary domains of engineering. For the near future, at least, this information will have to be requested from the human designer—or otherwise the system must be tailored to handle specific areas of application.

## An Expert System for Axiomatics

The ability to state the Axioms as clauses in a logical programming language allows for the development of an expert system for design axiomatics using Prolog.[11] The system architecture for an Axiomatic Advisor is compatible with the generalized structure shown in Figure A.7 of Appendix A.

The Kernel Module is a domain-independent component that is valid for diverse realms of design. The heart of the system is the Kernel Reasoner, which contains the Design Axioms and supporting functions such as procedures for determining functional independence and comparing information measures.

The user interacts with the overall system through the User Interface. This Interface provides a friendly demeanor so that the user may ignore the details of the system, such as the techniques used for knowledge representation. The Converser maintains a dialogue with the user, an interchange that may consist of a stream of questions and answers, or a series of menu-driven queries. The Artist is a graphic interface that presents the user with a pictorial representation of the knowledge encoded in the system.

The Administrator ensures that new knowledge is consistent with that already encoded in the Application Module. It keeps track of the identity and characteristics of different objects, as well as their mutual interactions.

The Document Generator, when invoked, prepares reports describing the encoded knowledge. The Kernel Gateway is an intermediary that mediates communications with the Application Module.

The Application Module contains knowledge that is specific to different realms of design. For example, a factory planning package should know that mechanical assembly equipment is not relevant to a polymer processing plant. Such knowledge is encoded in the Application Reasoner.

The Case Base contains knowledge of the particular design problem at hand, such as the functional requirements and constraints for a particular production plant. The Solution Board, a component of the Case Base, keeps track of the evolving structure or solution. The Solution Board provides the User Interface and Document Generator with the data relating to a specific factory design. The Application Gateway is the counterpart of the Kernel Gateway; the former mediates communications with the Kernel Module.

The Axiomatic Advisor package is currently under development at MIT. Further discussion of this architecture is available in the literature.[12]

## Implications for Design Automation

An automated designer must be guided by decision rules that limit the search space in intelligent ways. For such methods, a dichotomy exists in terms of generality versus efficiency: a general method applies to many domains of application, but is likely to be less efficient than a more specific method.

Under these circumstances, a careful trade-off must be effected to achieve a useful balance between the two conflicting factors. One proposed solution is the axiomatic approach to design.

The long-term goal of the axiomatics effort is to develop a design advisor that can dispense axiomatic reasoning in any area of design. An important lesson from the 1960's is that general problem-solving methods are weak due to the inability to manage the combinatorial explosion of the search space. The best-known example of this may be the General Problem Solver.[13]

For this reason, the most successful AI products have been those that pertain to highly specialized domains such as the diagnosis of bacterial infections. Does this imply, then, that the Design Axioms cannot be used to address the general design problem?

Not necessarily. The general design consultant described above incorporates a specific methodology—in particular, the axiomatic approach to design. As such, the system is already focused. In addition, the design philosophy of the automated designer freely permits the specialization of the Application Module to any level of detail. To illustrate, the Application Base for a system to design microprocessor-based devices may be decomposed hierarchically into consumer goods versus capital equipment; then a subordinate node of the latter category might deal with mobile versus stationary devices, and so on down to specific machines such as robotic manipulators. In this way, the dynamic knowledge base may be extended as needed to accommodate the use of the Advisor in novel fields of application.

An automated designer represents the ultimate goal of generic design methodologies such as the axiomatic approach to design. However, such an expert system will probably have to await the development of techniques for self-learning from formatted input such as textbooks, methods for learning from unstructured sources such as operational experience, knowledge representation schemes to encapsulate knowledge in disparate fields, and methods to reason about space and time. These areas represent current programs of research at various institutions around the world.

## Summary

The Design Axioms are generalized decision rules that apply to the synthesis of devices, both intelligent and otherwise. By serving as guidelines for evaluating alternative designs, they streamline the design process.

The Design Axioms may be stated with rigor through the use of symbolic logic. The development of mathematical expressions for the Design Axioms implies two types of consequences—theoretical and operational.[14]

On the theoretical side, symbolic logic has provided a tool for determining the precise relationships among the axioms and their derivatives. In particular, statements that were previously believed to be corollaries may actually be partitioned into distinct classes. One proposition is seen to be an immediate consequence of Axiom 1; in fact, it is an alternative statement of the axiom in an informal sense, but is identical in a formal sense. Most other corollaries are shown to be derivatives of the axioms plus certain reasonable assumptions.

On the operational side, the precise statements facilitate encoding the axioms in a programming language such as Prolog. The software statements then serve as the basis for a consultive expert system for design.

# APPENDIX D

# A Formal Framework
# for Learning Systems

*Love of wisdom without love of learning degenerates into utter lack of
principle. . . . Love of uprightness without love of learning degenerates into harshness. . . . Love of courage without love of learning
degenerates into mere recklessness.*[1]

Confucius

Learning refers to the modification of behavior patterns over time. Such adaptive
behavior is useful for dealing with unknown factors in current or future activities.
More specifically, they are useful in applications where (a) predictive knowledge of
the underlying process is incomplete, and (b) variations in inputs or environmental
conditions may exceed the predetermined range of values, or may even assume
wholly unexpected characteristics.

A basic intelligent machine is one that exhibits a range of behaviors, each action
corresponding to specific input stimuli. The mapping from input to output is determined by a human programmer and remains fixed unless respecified by human
intervention.

More sophisticated systems, however, must incorporate autonomous learning
behavior. By learning from experience, an automated system can modify its own
repertoire of behaviors over time. The advantages of an autonomous learning capability relate to the following factors:

- *Unavailable knowledge.* The knowledge required for effective performance is
  often unavailable. For example, the final thickness of a composite material
  emerging from a curing process depends on the history of temperature and
  pressure variables in ways which are still unclear to production engineers.
- *Elusive knowledge.* Sometimes a person may possess expertise, but cannot
  explain how he knows. A mundane example is a driver who makes the
  decision, based on "gut" feeling, to overtake a slow car in the face of a truck
  approaching from the opposite direction. A precise mathematical model of
  this situation would require differential equations, a methodology unknown to
  most drivers and not always recommended in an emergency.
- *Transfer bottleneck.* Even when an expert knows, and knows how she knows,
  it is often a laborious task to extract and codify the knowledge for translation

in software. Any number of examples can be found from current efforts to develop expert systems in engineering, business, science, and other fields.

These difficulties underscore the advantages of developing autonomously learning systems.

Learning systems may be categorized in terms of explicit versus implicit representation of knowledge. Explicit knowledge representation refers to the encoding of domain knowledge in terms and concepts readily comprehensible to the user. An example of explicit representation is the learning classifier approach.[2] Explicit knowledge may be declarative, as in the statement of facts, or procedural, as in the encoding of instructions.

In contrast, implicit techniques refer to the generation of appropriate output based on input stimuli, without the direct representation of knowledge for processing the information. A popular implicit approach is found in connectionist models. In this scheme, information processing know-how is encoded in terms of multiplicative weights at nodes and on the arcs between pairs of nodes. A key advantage of the neural approach is that the system itself performs the bulk of the work for establishing the implicit knowledge. A second advantage lies in robustness, in that a neural network will degrade gracefully rather than catastrophically when its structure is modified in minor ways.

The explicit and implicit methodologies are described more fully below, following an integrated framework for modeling learning systems in areas such as robotics and production control.[3]

## General Framework for Learning Systems

A *learning system S* is an entity that operates in some *environment E*. The object $S$ receives certain *input* stimuli $\Sigma$ from its environment. The domain of $\Sigma$ is a set of physical quantities such as pressure or temperature; its range is usually some subset of the real numbers.

The general model is depicted in Figure D.1.[4] The system $S$, as an adaptive unit, must incorporate mechanisms that specify its own behavior patterns. Hence $S$ encompasses the input transition function $M$ and the output function $N$, as well as the set $F$ of final states plus the other internal states of the kernel system $S_k$.

The kernel system $S_k$ encompasses the set $F$ of final states and some data $\mathbf{D}$. The terminal cluster $F$ is specified by the functional requirements $\mathbf{F}$, while the data $\mathbf{D}$ is determined by the messages from the input function $M$.

The main component of the input transition function $M$ is the set of input transition rules $R_M$. The function $M$ also encompasses the meta-level operators $\mathbf{O}$ and the plans $\mathbf{T}$ of the adaptive system.

The output function $N$ is specified by the rules $R_N$ which determine the external behavior of the system. These ideas may be encapsulated in the following algebraic model:

**Definition.** The structure $\mathbf{E} = \langle \mathbf{S}, \mathbf{F}, \Sigma, \Omega, M, N, F \rangle$ defines a finite deterministic *learning* automaton $S$ iff:

1. **S**, called the *state space* of *S*, is a finite nonempty set with $n > 1$ elements. Every element of **S** represents an internal state of *S*.
2. **F**, called the *functional requirement*, is a nonempty set of symbols. Each symbol specifies a goal state of *S*.
3. $\Sigma$ is a set of *input symbols*.
4. $\Omega$ is a set of *output symbols*.
5. The function $M : \mathbf{S} \times \Sigma^* \to \mathbf{S}$, is called the *complete transition* (or *next state*) function. If $s \in \mathbf{S}$ and $x \in \Sigma^*$, then $M(s,x) \in \mathbf{S}$ denotes the state that the automaton *S* goes to when *S*, while in state *s*, receives input *x*.
6. The function $N : \mathbf{S} \times \Sigma^* \to \Omega^*$ is called the *output*.
7. *F* is the set of *final states* defined by the functional requirements **F**.

In the definition above, $\Sigma^*$ refers to the collection of all concatenations (ordered sets) of symbols in $\Sigma$; and similarly for $\Omega^*$ as the concatenations of symbols in $\Omega$. These concepts may be further illustrated by a simple example. Consider a robot whose task is to assemble an automotive engine. The functional requirements **F** might be encoded as the final states *F* in terms of a photographic image of the final assembly as well as physical characteristics such as low friction in the crankshaft. Input sensors such as vision and touch provide information $\Sigma$ which may be stored as data **D**.

The transition function *M* may regulate activities such as the sequence of assembly operations. The rules base $R_M$ contains knowledge such as "If the piston rings are in place, initiate the piston-crankshaft assembly process," or "If a piston shows discoloration, check for structural integrity." The set of plans **T** determines how the rule base $R_M$ may be improved by selecting appropriate operators from the operator

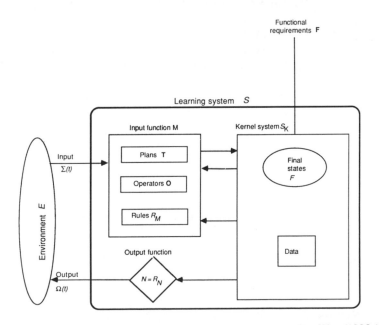

**Figure D.1** General framework for self-learning systems. (After Kim (1988e).)

base **O**. The output function $N$ would contain knowledge such as the manipulator motions required to insert pistons into place.

The deterministic automaton $S$ models a device that changes state in predictable fashion on the basis of a particular input. This situation ensues when the transition rules are functions that exhibit unique domain values (i.e., "conditions," "antecedents," or "detectors"). This implies, for example, that no two decision rules in the rule base $R_M$ have the same configuration of detectors.

In contrast, when two decision rules can respond to the same input (i.e., they have the same detector), then the device is nondeterministic. Further, if the competing rules may be assigned specific probabilities of being selected, then the system is a probabilistic automaton.

Many real-world processes tend to exhibit random characteristics. This phenomenon gives rise, for example, to problems with statistical quality assurance in manufacturing environments.[5]

In such stochastic domains, the automata-theoretic model may be generalized to include probability distributions over state transitions.[6] These models can be used as the basis for the determination of stochastic behavior such as mean transition times and rates of adaptation.[7] However, the conventional viewpoint in stochastic studies focuses on analytic techniques and does not seem to serve as well for the investigation of fundamental capabilities and limitations of intelligent systems.[8] For these reasons, algebraic structures such as the automata framework are particularly suited for the study of fundamental issues.

## Application to Production Rule Systems

This section discusses the application of the general adaptive model to production rule systems. In this approach the knowledge is represented explicitly as a set of rules in terms of triggering conditions and corresponding actions. A specific category of the production approach is found in learning classifier systems.

### Structure

In a production rule system, the system $S$ operates by utilizing a self-learning strategy or adaptive *plan* **T** in order to attain a set of *functional requirements* **F** (see Figure D.2). In a composites processing application, **F** might have two components relating to density and thickness in a cured part.

The degree to which the functional requirements are met is indexed by a *performance measure* $\pi$. If a binary range for $\pi$ is chosen, for example, then we might use $\pi = 1$ if all the functional requirements are met, and $\pi = 0$ otherwise.

The domain of action of the adaptive plan **T** is the set of *rules* **A**. **A** may be interpreted as the potential knowledge base of decision rules that $S$ may employ. In general, the knowledge base **A** is *potential* rather than *actual*: at any time $t$, $S$ may have only a small subset of the possible set of rules.

The adaptive plan **T** yields a sequence of selections from a set of *operators* **O**. In other words, **T** invokes particular operators $o$ in **O** to modify structures in **A**. An

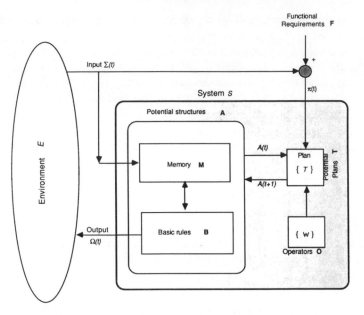

**Figure D.2** Framework for production-rule systems.

example of an operator is the splicing and mixing of two rules in **A**; this operator, known as a *genetic algorithm*, is discussed further in a later section.

The set of general structures **A** may be partitioned into two subsets: the set of *basic structures* **B** and the *memory* **M**. We use the formalism **A** = ⟨ **B**, **M** ⟩ to indicate that **A** is the ordered pair consisting of **B** and **M**. For some basic rule $B$ in **B**, let $B(t)$ denote the particular item applied at time $t$ in order to fulfill the functional requirements **F**. At each cycle, the basic rules **B** observe the contents of memory **M** and may choose to generate a message. This message may be an output $\Omega$ directed to the environment or a memo $m$ sent to augment the memory **M**.

### Operation

In general, the adaptive plan **T** is a stochastic rather than deterministic process. Instead of a predetermined structure $A(t+1)$ obtained from $A(t)$ and $\Sigma(t)$, **T** assigns probabilities ⟨ $P_1$, $P_2$, . . . ⟩ to the permissible set of rules ⟨ $A_1$, $A_2$, . . . ⟩ and chooses the next rule to activate.

Let **P** be the set of permissible probability distributions over **A**. Then **T** first selects $P(t+1)$ from **P**, followed by a selection of $A(t+1)$ from **A**. For the deterministic case, $P(t+1)$ assigns the single value 1 to a particular rule $A$ in **A**.

In practice, the rule $A(t+1)$ is obtained from $A(t)$ by using an operator from the set **O** = { $o : \mathbf{A} \to \mathbf{P}$ }. The adaptive plan is then given by the mapping **T''** which translates the input and rule set into the output. In other words, $\mathbf{T'} : \Sigma \times \mathbf{A} \to \Omega$.

*Comparison of Adaptive Plans.* A key requirement in using self-learning systems is to determine effective adaptive plans under differing environmental conditions. Let **T''** be the set of adaptive *plans to be compared* for use in the set of possible environments **E**.

For a specific environment $E$ in **E**, the input $\Sigma$ to the plan must contain some measure of the efficacy of **T** in fulfilling the functional requirements **F**. In other words, a component of $\Sigma(t)$ must contain the payoff $\pi(A(t))$, given by the mapping **A** $\rightarrow$ *Reals*, which measures the degree to which the functional requirements are met in the environment $E$.

In the special case where the adaptive plan acts only on a knowledge of the reward, $\Sigma(t) = \pi_E(A(t))$. Plans that utilize information in addition to reward should perform at least as well as those that use only payoff.

To compare the efficiency of adaptive plans against each other, we must define some criterion $k$. Constructing such a criterion is a nontrivial task. For example, consider the following two plans: one performs superbly in some environments but poorly in others, while the other performs reasonably well in all cases. In this case, the selection of one plan over another should be guided by the particular application.

*Systems Without Adaptive Capability.* The formal framework discussed in the preceding sections may be tailored for particular domains. For example, it may be applied to intelligent manufacturing systems whose capabilities remain constant over time. Since learning is absent in such a system, the memory set **M** consists of the following: a singleton $\Sigma(t)$ at time $t$, and the empty set for the memo bank **m**. $\Sigma(t)$ may, of course, still incorporate payoff information $\pi(t)$ which measures how well the functional requirements are satisfied.

The basic rule set **B** is static: it remains unchanged from one time instant to the next. Hence the "adaptive" plan **T** is a degenerate mapping from $\Sigma \times$ **A** $\rightarrow$ **A**. In other words, the rule at time $t$ is given by:

$$A(t) = \langle B, M(t) \rangle = \langle B, \Sigma(t) \rangle$$

The second equality follows from the fact that the memory $M$ is a degenerate function of the input $\Sigma$.

## Interpretation for Classifier Systems

*Structure.* A category of rule-based systems that has been studied extensively is that of *learning classifier systems*.[9] The classifier technique has been applied to a variety of simplified domains, from negotiating mazes[10] to controlling simulated pipelines.[11]

A classifier system is a program that learns rules called *classifiers* which define the system's response to some environment. The two central features of a classifier system are:

- Rule and Message System.
- Credit Assignment System.

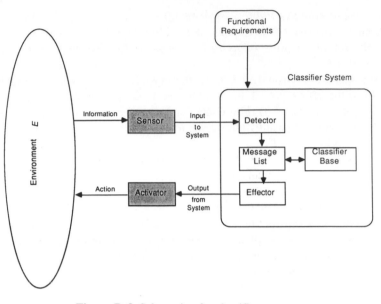

**Figure D.3** Schematic of a classifier system.

A schematic diagram is given in Figure D.3. A popular method of assigning credit and modifying the rule base is the genetic algorithm.

The *detector* subsystem receives input concerning the state of the environment; the input information is decoded according to a prescribed *message* format. This message is placed in a common area called a *memo bank, blackboard,* or *message list,* together with other internal messages remaining from the previous cycles. Messages on the message list may stimulate rules called *classifiers*, stored in the *classifier base*.

An activated classifier may be selected to send a message to the message list for the next cycle. In addition, other messages may specify external action through a set of rules called *effectors*.

*Operation.* In conventional rule-based systems, the value of a rule is specified by the system developer. In a learning classifier system, however, the relative utility of each rule is a vital item of information that must be determined by the system itself.

Such learning is facilitated by the context of the competitive economy in which classifiers bid for the right to respond to relevant messages. The highest bidders are awarded the contract; if they are successful in fulfilling their goals, then they are rewarded with an income that may be used for bidding at the next cycle. In such a competitive economy, the fittest rules survive and the unfit eventually die away.

Biological species survive dramatic changes in the environment by mutation and the juxtaposition of different genes. Similarly, a classifier system provides for the generation of new rules through the mechanism known as the genetic algorithm.

This algorithm involves the generation of new rules through the mutation of old ones. One mechanism for mutation is the interchange of the components of two rules (analogous to chromosomes) after splicing at a random position. For example, consider two strings $S_1 = \langle B, A, ?, B, A \rangle$ and $S_2 = \langle A, B, B, A, ? \rangle$, where the question mark indicates that the identity of the component or gene is immaterial. If the random position selected is 2, then one offspring $T_1$ obtains the first 2 genes of $S_1$ and the last 3 genes of $S_2$; that is, $T_1 = \langle B, A, B, A, ? \rangle$. Conversely, the other offspring is $T_2 = \langle A, B, ?, B, A \rangle$.

## Application to Connectionist Networks

This section interprets the general adaptive model in the context of neural nets, also known as the connectionist model. For this model, we may specialize the general framework to the neural model shown in Figure D.4.

In general, the connectionist approach does not store input data **D** in explicit form. Rather, the inputs are absorbed into the organizational structure of the system. More specifically, the input is transformed into a set of node values plus arc weights that depict the relationships among the nodes.

The transformation rules are procedures that specify how the data is to be converted as a function of the input and the current system state. These rules are defined by the weighting strategies as well as the parameters used; an example is

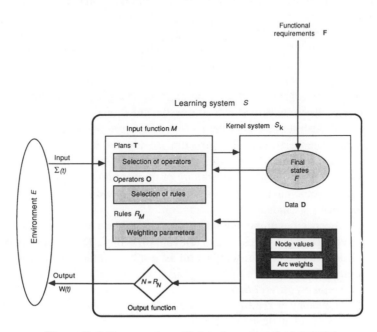

**Figure D.4** Framework applied to the connectionist model.

found in the "momentum" factors which specify how much influence the new input is to have on the existing state.

The operators identify the rules to activate or modify. Current connectionist models tend to use a null operator set; that is, the initial rules for a system remain permanently active and are unchanged for the entire simulation. The plans pertain to the selection of a specific set of operators for a particular combination of system and environment.

The general connectionist model may itself be specialized for specific applications. For example, the feedforward connectionist model[12] is a neural net in which the lateral arc weights may be considered null (i.e., having zero value.)

## *Discussion*

The obvious advantage of explicit over implicit knowledge representation is the ready interpretation of the newly acquired knowledge. Ideally, the system should be able to inform the user of its findings, not merely transform a set of inputs into effective outputs.

Unfortunately, existing connectionist techniques do not allow for this explanatory facility. In the future, however, we might take a cue from the human brain (as well as the levels of computer languages, from machine language to high level specifications.) At lower levels of representation in the human brain, data is encoded in the form of voltage differences and structural changes, both mediated by chemical action. Yet, at some level of awareness, we are conscious of aggregate concepts (e.g., systems), relationships, (e.g., bondings), and abstractions (e.g., freedom).[13]

## Software Structure

Since the software organization must implement the generic functionality of a learning system, its architecture will mirror the functional organization given in Figure D.1. To illustrate, consider the software structure of a classifier system. A central design issue pertains to the structure of the set **B** of basic rules. Each decision rule or *generalized chromosome* in the basic set **B** must consist of a detector, effector, and strength. The *strength* of the chromosome is a measure of its utility. A chromosome having a high strength value should be retained, while one with a low value is a candidate for elimination.

The *detector* specifies the conditions under which the chromosome may "fire" or be activated. The *effector* specifies the set of actions to be taken when the chromosome is activated. The components of each detector or effector may be called *generalized genes*. Each gene may take one of a set of values; each such value is called a *generalized allele*.

These concepts may be readily expressed in Pascal-like statements. Let detectorAlleleType be the set of allele values that may be assigned to each detector gene; there are MAXDETECTOR such genes in a detector. Similarly, let effectorAlleleType be the set of alleles for MAXEFFECTOR genes in the effector. Then the detectors and effectors may be prescribed to be instances of the following types:

detectorGeneType : detectorAlleleType;
effectorGeneType : effectorAlleleType;
detectorType = Array[1. . MAXDETECTOR] Of detectorGeneType;
effectorType = Array[1 . . MAXEFFECTOR] Of effectorGeneType;

For example, each element of the detectorType array is a gene whose value is taken from the pool of values in detectorAlleleType.

Using these definitions, we may specify each chromosome as:

chromosomeType = Record
    strength : Real;
    detector : detectorType;
    effector : effectorType;
            End;

In particular, strength is a variable that can accept a real value. A representative range for this parameter is the set of real numbers from a low of 0 to a high of 1.

If there are MAXCHROMOSOME number of chromosomes in the learning system, its genetic library is of the following type:

chromosomeBaseType = Array[1 .. MAXCHROMOSOME] Of
    chromosomeType;

This final structure corresponds to the set **B** of basic structures depicted in Figure D.2. These concepts are clarified further through examples in the next section.

## Illustrative Applications

### *Mobile Robot*

The concepts introduced in the preceding section may be clarified through application to a simple context pertaining to a mobile robot. Suppose that a robot $R$ is placed in an amusement park with the purpose of picking up litter such as empty soda cans; the robot might just as easily seek out inoperative machine tools in an automated factory. We assume the following:

1. $R$ can distinguish eight directions in its immediate neighborhood: north ($N$), northeast ($NE$), east ($E$), southeast ($SE$), south ($S$), southwest ($SW$), west ($W$), and northwest ($NW$). Each of these regions in $R$'s immediate neighborhood is called a *cell*.
2. $R$ can recognize three types of objects in any cell: open space ($S$), obstruction ($O$), and litter ($L$). An example of an obstruction is a person, a cart, or a pillar; an example of litter is a stray popcorn container.
3. Each cell contains exactly one object: $S$, $O$, or $L$.
4. At each cycle, $R$ moves into an adjacent cell.

5. *R* does not move into any cell containing an obstruction.
6. *R* succeeds when it moves into a cell containing litter (which it recovers during the move).

In this scenario, the learning system is the robot *R*, having the functional requirement of gathering litter. The detector component of each chromosome consists of eight generalized genes corresponding to the eight directions; each gene can be expressed as one of three alleles (space, obstruction, or litter). The effector for each chromosome consists of a single gene whose alleles are the eight directions.

An example of a chromosome might be

$$S, ?, O, S, S, S, L, S;$$
$$W;$$
$$0.5$$

The interpretation is: "If there is a space (*S*) in the north cell, anything (?) in the northeast, an obstruction (*O*) to the east, litter (*L*) to the west, and space (*S*) everywhere else, then go west (*W*)." The strength of this rule is 0.5 on a scale from 0 to 1.

The robot may begin with a random collection of such chromosomes. By rewarding chromosomes that lead to successful behavior and penalizing others, *R* will eventually accumulate an effective set of rules.

This is particularly true when the environment exhibits some orderly—rather than purely random—behavior. For example, suppose that litter is to be found to the north of every pair of blocking objects. The robot will ultimately behave in a way that takes account of this regularity in the environment.

## Production Process Control

This section illustrates the adaptive framework by describing it in the context of production control applications such as composites fabrication. Composites are used in a wide variety of structures; an illustrative application is found in advanced aircraft bodies where they are used for their superior strength-to-weight ratios.

Although standards exist for many composites, their characteristics vary from lot to lot even when produced by a single manufacturer. Because the chemical processes and curing characteristics are not fully understood, final part quality may differ significantly due to factors such as the existence of voids and differing chemical structures in the final cured part.

The control strategies to date have usually relied on simple open-loop processes. However, a more effective methodology is to incorporate the following features: closed-loop control, real-time sensing and activation of effectors, and self-learning techniques.

Figure D.5 depicts an improved composites curing scheme. The process is designed to take material input in the form of fibers and resin, and to produce composites parts with specific functional requirements such as thickness *H*, fiber volume fraction $V_f$ (i.e., volume of fibers as a fraction of the total part volume), or uniform resin density *r*. The functional requirements are to be met by design

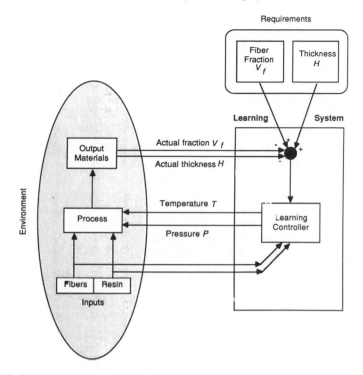

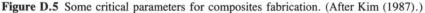

**Figure D.5** Some critical parameters for composites fabrication. (After Kim (1987).)

parameters such as the mold temperature $T$ and pressure $P$. The penalty function is determined by the difference between actual and desired values of the requirements. Its mathematical structure will depend, of course, on the particular application.

## Summary

A learning system, as an intelligent entity, can be examined through the factors of intelligence. Three significant parameters for a learning system are function, software, and hardware.

This appendix has explored an algebraic model of learning systems. The model was used as an abstract foundation for the key parameters, and the way they relate to specific application domains.

Many questions remain open in the development of self-learning systems. Examples of such issues are the following:

- How robust is system performance as a function of hardware and software resources? Incorporated in the general model are implicit assumptions such as the adequacy of storage to reflect a rich behavior space, or the speed of learning to accommodate rapid changes in the environment. Does performance degrade gracefully when the underlying assumptions are violated?

- What are the trade-offs between uniformity and efficiency in information representation? What data structures and models are appropriate? Suppose, for example, that a relational database model is to be used for one component of the knowledge base. What are the relative merits of using a logical programming language such as Prolog versus a procedural language such as C? For which modules does the computational efficiency of a procedural language supercede the advantages of a declarative approach?
- What are appropriate data reduction techniques for consolidating realtime information obtained from the sensors? What is the optimal nature and degree of data reduction at each stage? For example, should a microprocessor in a transducer transmit only a moving average of sensor readings, to counteract the effects of noise? When are filtered or compressed statistics, such as the median or standard deviation, appropriate?

Issues such as these deserve detailed study. By addressing them in a systematic fashion, we may prepare a solid foundation on which to build learning systems of increasing sophistication.

# APPENDIX E

# A General Model
# of Information

*Fundamental ideas play the most essential role in forming a physical theory. Books on physics are full of mathematical formulae. But thought and ideas, not formulae, are the beginning of every physical theory. The ideas must later take the mathematical form of a quantitative theory, to make possible the comparison with experiment.*[1]

Albert Einstein

Information is knowledge that relates to the condition of a system or its environment. In a universe of perfect predictability, a system would not need to monitor information while pursuing its goals. Since each condition could be foreseen, a system could perhaps be programmed appropriately at the outset and relied upon to perform as specified.

On the other hand, the complete modeling of an environment of even moderate complexity will likely tax the memory and computational resources of a system. Consequently, any realistic system will possess only partial knowledge of the world, and will have to deal with unforeseen circumstances as they arise.

A general model of information may be based on the probability of attaining a set of objectives. This perspective is a generalization of the classical model of information used in the field of communication engineering.

This appendix introduces the classical theory of information and examines its limitations. Then it presents a generalized information model that is used to explore some information characteristics and their implications for system performance.

The classical viewpoint is shown to be a special case of the general model, and a number of critical information characteristics are discussed. The presentation is based on several previous publications.[2]

## Classical Information Theory

The classical model of information was developed for use in the field of communications engineering. This formulation depends on the probabilistic nature of a predefined set of symbols that must be transmitted from a sender to a receiver.

More specifically, let $\{s_1, s_2, \ldots s_n\}$ be a set of symbols that may be transmitted. Any symbol $s_i$ might be sent from the source with probability $p_i$. If the likelihood of sending symbol $s_i$ is small, then the informational value of receiving that symbol will be high. Hence the quantity of information should be inversely related to the probability of receiving any particular message. In other words, the information $I_i$ due to receiving message $s_i$ will be inversely proportional to the probability $p_i$.

The specific relationship between $I_i$ and $p_i$ should be consistent with various intuitive notions of information. For example, the overall likelihood of a sequence of two independent symbols being sent is the product of the individual probabilities, namely, $p_i p_j$; but the overall information should be the sum of the respective information values, namely, $I_i + I_j$. A relationship[3] that satisfies the necessary conditions is given by

$$I_i = \text{lb} \ (1/p_i)$$

where lb denotes the binary logarithm, or log to base 2.

With this relationship for $I_i$, the mean value of information transmitted over a communication channel is given by

$$I = \Sigma_{i=1}^m \ p_i \cdot \text{lb}(1/p_i) = -\Sigma_{i=1}^m \ p_i \cdot \text{lb}(p_i)$$

This formula has been used with remarkable efficacy in communication engineering and has also found some application in other fields ranging from biology to psychology and economics.[4]

The classical information model, however, has its limitations. By focusing only on the statistical properties and ignoring the semantic content of the symbols, we are left with a metric that is of limited use in modeling decision-making situations.

For example, suppose that a production manager receives a report stating that the following year's sales forecast $s_i$ for the Elixir of Youth is 59.3 tons per day. What is the "information content" of this communication? There are major difficulties in attempting to apply classical information theory in this context.

- *Unknown probability distribution.* The classical theory assumes that $p_i$, the probability of occurrence of symbol $s_i$, is well defined. In our example, one would be hard-pressed to offer a probability density *a priori*.
- *Lack of a predefined set of symbols or messages.* The classical definition of information is based on a predefined, finite set of symbols $\{s_i\}$ originating from an information source. In our example, the set of messages corresponds to the set of nonnegative real numbers; defining a probability density for an uncountably infinite number of outcomes can sometimes be troublesome.
- *Blindness to semantic content.* In addition to the technical difficulties above, the classical definition of information was never intended to capture semantic content or value. A string of bits representing a sales forecast is calculated to have the same "information" whether received by a person, a cuckoo clock, or no one at all. This is somewhat at odds with our everyday notion of "information."

- *Novel situations.* Any intelligent system of reasonable complexity will need to respond to novel environmental conditions. The very nature of these novel developments may not be known in advance. But it is precisely such novel, unanticipated developments that are most critical to the attainment of system objectives.

For these reasons, the classical notion of information must be generalized for use in decision-making environments.

## General Information Model

The general model of information is based on a cybernetic perspective that takes into account the effect of a message on a system's ability to achieve its goals.[5] Consider a purposive agent with a goal set $G$, consisting of one or more objectives. Let $E$ be the event associated with the attainment of goal $G$, and $p$ the probability that $E$ occurs. Let $p_0$ be the prior probability before the receipt of a communication $C$, and $p_1$ the posterior probability. The information content of $C$ is defined as

$$I = \mathrm{lb}(p_1/p_0)$$

In other words, information is a measure of the probability of success. A useful communication will increase the likelihood of attaining success; hence $p_1 > p_0$ and $I > 0$. On the other hand, if a communication reduces the likelihood that $E$ will occur, then $p_1 < p_0$ and therefore $I < 0$. Such a communication conveys disinformation.

By writing the information equation as $I = \mathrm{lb}(1/p_0)-\mathrm{lb}(1/p_1)$, we see that information is the logarithmic difference in the inverses of probabilities. This interpretation is consistent with the proposition that the information needed by a subsystem to perform a given task is equal to the difference in the information required for the task, versus that which is already available.[6]

A number of immediate corollaries are as follows.

- If a communication does not affect the probability of success, then the information content of the message is 0. This is clear from the following observation: if the probability of success is unaffected, then $p_1 = p_0$. Hence $I = \mathrm{lb}(1) = 0$.
- If the probability of attaining the goal state is 1, then no communication can convey positive information. Since $p_0 = 1$, the information content is $I = \mathrm{lb}(p_1) \leq 0$. Here a communication can convey no information at best, and disinformation otherwise.

## Classical Theory in Terms of the General Model

Classical information theory is the offspring of communication engineering, a discipline dealing with the goal of a communication channel to reproduce messages faithfully. Let $\{X_1, \ldots, X_m\}$ be the set of messages or symbols originating at a source $S$, with probability set $\{p_1, \ldots, p_m\}$. These symbols are transmitted by a channel to a destination $D$. Upon receipt of a message, the agent at the destination must determine which of the $m$ messages it is.

Suppose the agent is asked to select among the $m$ messages before having received any information. In this case the decision maker might as well select randomly among the messages. If the actual message to be sent is $X_i$, then the prior probability that the agent will successfully select $X_i$ is $p_i$. After receipt of the communication, the probability of success is 1. Since the prior and posterior probabilities are $p_i$ and 1, respectively, the information content of the communication is

$$I_i = \text{lb}(1/p_i)$$

The average information or variability among all $m$ messages is then

$$I = \Sigma_{i=1}^{m} p_i \cdot \text{lb}(1/p_i)$$

which is the classical formulation of information. Hence the classical model may be viewed as a special case of the generalized information model.

## Characteristics of Information

Descriptions of physical systems, being idealized models, will often repesent reality imprecisely. For example, a physical plant will be subject to disturbances, not all of which can be accounted for in advance.[7] Hence our confidence in any component designed to control a physical plant will increase if realtime data on both the model and plant are monitored and reconciled.

In a static situation, process information is unnecessary. The system can be configured at the beginning and trusted to behave as originally implemented. In reality, the system may change (e.g., due to wear and tear) as may the environment (e.g., as reflected in temperature fluctuations). For these reasons, the system requires information to cope with disturbances.

Even when information about a situation is available, however, it may be misleading by being irrelevant (not effective), incomplete (not sufficient), or delayed (not timely). Usually, however, the controller does not require complete information about a situation in order to make proper decisions. Visual information impinging on the human retina, for example, is aggregated and condensed through several levels until only a shadow of the original diversity filters into the visual cortex. Since much of the input data is redundant, however, the reduction in classical information may still leave the generalized information largely unscathed.

### Sensitivity

The *sensitivity* of information may be defined as the elasticity of performance with respect to the information variable. Let $P$ denote a performance index relating to a functional requirement. An example of a performance index might be the maximum power output of an automotive engine.

Further, let $I$ denote the classical measure of information. As discussed previously, the classical metric monitors the variability in a stream of messages, while ignoring the goals of an adaptive system. Hence the performance of an agent may or may not be affected by the amount of information in classical terms.

We define the *sensitivity* of an information stream by the fractional change in performance, $\partial P/P$, resulting in a fractional change in information, $\partial I/I$:

$$S = \frac{\partial P/P}{\partial I/I}$$

where $\partial$ denotes the partial derivative.

By the property of logarithms, $S$ is also given by the following relationship:

$$S = \frac{\partial \ln P}{\partial \ln I}$$

where ln denotes the natural logarithm. The relationship between high and low sensitivity is depicted in Figure E.1.

Let $I_0$ correspond to the minimal information needed to yield the performance level $P_0$. For large values of ln $I$, we would expect diminishing returns in ln $P$. Hence the ln $P$ versus ln $I$ curve will level off.

If the decision-making entity must process the stream of incoming information to sift out the critical components, excessive information input may well lead to decreased performance. Since a positive variable is monotonically related to its logarithm, we may infer from Figure E.2 that $P$ begins to decrease for $I > I_1$.

Such a declining curve reflects the phenomenon of *information overload* often observed in biological subsystems. At a higher systemic level, the curve mirrors the proposition that decision makers often receive too much irrelevant information rather than too little in a highly computerized society.

If we assume that the acquisition and/or processing of information requires

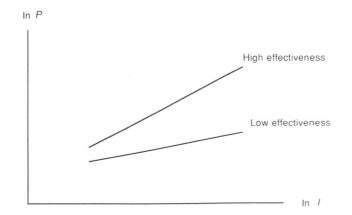

**Figure E.1** The relative effectiveness of different types of information.

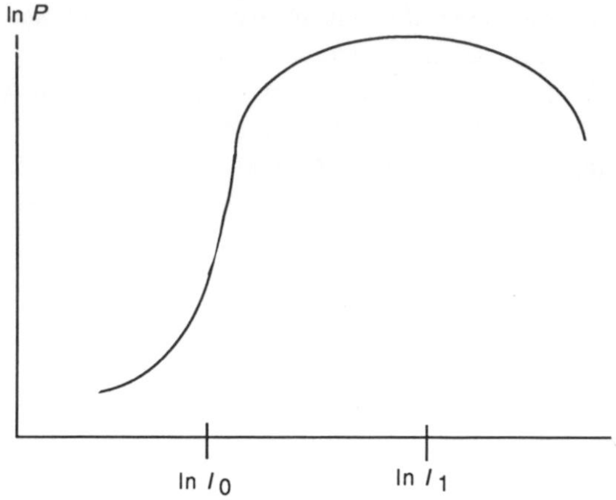

**Figure E.2** A rising-falling performance curve as a function of information.

valuable resources, then we may infer that a system should avoid information requirements beyond the minimum required to attain its goals. In terms of Figure E.2, the principle of information minimization would require that the actual information $I''$ be as close as possible to $I_0$ without falling below this threshold. This idea is better addressed by the criterion of efficiency.

## Efficiency

*Efficiency* is a measure of the parsimony with which useful information is conveyed. Let $I_0$ be the *sufficient*, or minimal, amount of classical information required to fulfill a particular functional requirement. Then the actual information $I''$ may be less than, equal to, or greater than $I_0$.

- $I'' < I_0$. The actual information $I''$ is insufficient for the task intended. The *relative deficiency* $\delta = (I_0 - I'')/I_0$ is a measure of the extent to which the actual information falls short of the amount needed.
- $I'' = I_0$. The actual information is also the sufficient information. In this case the information minimization principle is satisfied.
- $I'' > I_0$. The actual information is in excess of the minimum required. The *efficiency* $\eta = I_0/I''$ is a measure of information parsimony. Its complement, $R = 1 - \eta$, may be called *redundancy*.

By construction, all three metrics, $\delta$, $\eta$, and $R$, take on values in the half-open interval $(0,1]$. Efficiency may fall below 1 when redundancy is required in order to guard against the effects of noise or the possibility of failure in various components.

## Discussion

A third attribute of information is *timeliness*, which denotes the extent to which the promptness or tardiness of a communication affects system performance. This characteristic was discussed in detail in Part II under the dimension of time.

The attributes of sensitivity, efficiency, and timeliness are convenient parameters for the purposes of discussion. However, these are not independent concepts since they are linked at least indirectly through the parameter of system performance.

Suppose, for example, that $s$ is the actual speed for a robot whose functional requirement is $s'' = 10$ km/hr. Then the sensitivity for $s$ is defined whether or not the actual design meets the requirement of 10 km/hr.

In contrast, efficiency and timeliness reflect the economy with which a design achieves the desired functional requirements. Hence they are defined only when the functional requirements are satisfied, and not otherwise. For example, it makes little sense to speak of the efficiency of performing a task when the requisite task is not even fulfilled.

The efficiency and timeliness of information are vital concerns in computer-based systems.[8] For example, the use of microprocessors results in information loss due to temporal lags and the discretization of continuous information. The increasing prevalence of computerized control and intelligent devices implies that such issues will become even more critical in the years ahead.

## Summary

A quantitative formulation of information may be based on the likelihood of attaining a set of goals. This cybernetic model of information, which encompasses the classical measure as a special case, serves as the basis for a number of critical metrics such as the sensitivity and efficiency of information.

# APPENDIX F

# Levels of
# Intelligent Design

*A rough division can be made of six types of occurrences in nature. The first type is human existence, body and mind. The second type includes all sorts of animal life, insects, the vertebrates, and other genera. . . . The third type includes all vegetable life. The fourth type consists of the single living cells. The fifth type consists of all large scale inorganic aggregates, on a scale comparable to the size of animal bodies, or larger. The sixth type is composed of the happenings on an infinitesimal scale, disclosed by the minute analysis of modern physics. Now all these functionings of Nature influence each other, require each other, and lead on to each other.*[1]

Alfred North Whitehead

Intelligent behavior can often be partitioned into layers for both analysis and synthesis. We have seen how this relates to biological systems, where physical signals are processed at the neural level and eventually funneled through a series of integrative mechanisms until the resulting information can be used for high-level reasoning. The converse process takes place when a conscious decision to reposition a limb moves through the layers of information-processing mechanisms, from the cerebral cortex down through the spinal cord, and eventually to motor neurons. In a similar way, the high-level task instructions in an application program are translated into assembly language, where the statements in turn are converted into machine language for eventual execution on physical devices.

The stratification of information processing into layers is also mirrored in the way intelligent systems can be used to design other systems. This appendix presents a layered, hierarchical framework for designing artifacts, beginning with the general notion of design and descending through automated design packages, followed by intelligent mechanisms for producing physical goods. For the sake of concreteness, the discussion involves the case study of an expert system to design flexible manufacturing plants which in turn produce mechanical devices.

Table F.1 presents a layered view of intelligent artifacts. Level 0 is the basic layer referring to the final product. This object may be as intelligent as an autonomous robot or as dull as a doormat.

The basic-level object is fabricated by a producer, whether human or otherwise.

*Table F.1* Levels of intelligence in system design. Some products are intelligent, and others are not; both are fabricated by a production system, which in turn might be designed by an expert program. The Meta Advisor contains knowledge engineering techniques for designing domain-based expert systems.

| Level | Description | Instances | Intelligence |
|-------|-------------|-----------|--------------|
| 0 | Product | Gears; calculators | Sometimes |
| 1 | Plant | Factory | Yes |
| 2 | Advisor | Factory configuring expert system | Yes |
| 3 | Meta Advisor | Intelligent system designer | Yes |

This producer, residing at Level 1, generally displays a repertoire of behaviors and can be considered intelligent.

The production plant may be designed by an expert system or program at Level 2. A producer must be designed in conjunction with the ultimate goods of fabrication; a factory for apparel, even if it is flexible, will differ from a comparable one for lamps. Hence, an expert system for plant design must contain knowledge of the ultimate product as well as facilities for its fabrication.

To carry this line of thought further, expert systems share commonalities in conception, structure, and construction procedures. Further, an expert program is itself an intelligent system that can be interpreted in terms of the six factors of intelligence. The relevant knowledge can be incorporated into a Meta Advisor, located at Level 3, to automate or assist in the design of domain-level expert systems.

The next section describes a knowledge-based system developed at the Massacusetts Institute of Technology. It is an expert system at Level 2, tailored to the design of production facilities for making mechanical gears.[2] This system, as well as its referent domain of production plants, is presented in the context of the factors of intelligence.

The appendix concludes with a discussion of current research activity dedicated to the development of a Meta Advisor at Level 3. This project involves a system architecture to incorporate knowledge engineering methodologies, project management tools, and intelligent system principles for the construction of basic expert programs and other smart artifacts.

## Factory Design Advisor

An expert system for factory design has been developed to demonstrate the utility of the six-dimensional framework for intelligent systems. In order to constrain the task to a manageable size, the software addresses the design of flexible factories for producing gears.

Since the advisor and the flexible factory are smart systems, both can be synthesized and analyzed in terms of the factors of intelligence. These dimensions and their attributes are highlighted in Table F.2. The first set of entries in the table indicates, for instance, that the production plant must be able to build from one to a thousand types of gears concurrently, in lot sizes ranging from a singleton up to one million units.

*Table F.2* The plant and advisor in terms of the factors of intelligence

| Factor | Plant (Level 1) | Advisor (Level 2) |
|---|---|---|
| Purpose | Manufacture from 1 to 1000 gears concurrently | Improve quality by 25% |
| | Accommodate lot sizes from 1 to 1,000,000 | Reduce design time by 90% |
| Space | Environment demands higher productivity, enhanced quality, increased diversity, shorter lead times | Environment: interfaces to user and data sources |
| | Technology takes the form of machine tools, sensors, inventory software, etc. | Technology: knowledge engineering techniques implemented on graphic workstations |
| Structure | Hardware in the form of tools, conveyors, local networks, etc. | Domain-independent inference modules and interfaces |
| | Software in computer-aided design/manufacturing, etc. | Domain-dependent knowledge of factories and case base |
| | Human programmers and others | |
| Time | Micro: sequencing of operations | Realtime interaction with user and access of databases |
| | Macro: throughputs, reconfiguration delays, equipment utilization, etc. | Evolution of capabilities |
| | | Learning through feedback on previous designs |
| Process | Coordination of multiple, decentralized processes | Forward reasoning from principles and backward from case goals |
| | Dynamically reconfigurable | Enhancement of input data |
| | Basically passive | Divide and conquer strategy |
| | | Coordination of realtime interaction and batch computations |
| Efficiency | Hardware: high utilization through versatility | Rapid inferences through heuristics |
| | Software: optimized procedures | Satisficing design, with optimization where feasible |
| | | Minimal memory through appropriate data structures |

The issues involved in designing a flexible factory were discussed at length in Chapter 12, and we will forego their reiteration here. The remainder of this section deals with the design of the expert program viewed in the light of the framework for intelligent systems.

## Purpose

The design of an expert system, as with other engineered products, begins with the perception of a need and its translation into a set of specifications. The purpose of a knowledge-based system is to liberate or assist a user in making decisions within a domain of application. By incorporating knowledge from diverse sources, the performance of an expert program can match or even exceed that of humans.

The Factory Configuring ExperT (FACET) assists in the design of flexible factories for manufacturing gears. Its functional requirements are to improve the quality of the designs by 25 percent, while serving to reduce the turnaround time for design by 90 percent.

The increase in the quality of designs may take several forms. One of these is to employ the proper mix of machine tools to manufacture gears of consistent dimensions. Another mode of enhancement is to design a factory having high utilization and thereby constrain long-term operating expenses. Additional objectives include the minimization of scarce factory space or the initial cost of plant construction. Still other objectives relate to the enhancement of reliability, serviceability, and availability.

These enhanced capabilities are possible through the integration of knowledge from a spectrum of sources, including multiple human experts and heuristics from design manuals. Another mechanism for quality enhancement is through a systematic exploration of alternative designs rather than a straightforward dash for the first satisfactory solution.

The reduction in turnaround time for designing a factory is achieved by automating most of the routine aspects of decision making. These tasks include the acquisition of supporting information from external data banks in response to the needs of the case. Another task for automation is the generation of reports, either in whole or in part, to finalize the results of a design session. On the other hand, some auxiliary functions, such as interfacing to inquiries from the managment, cannot be adequately computerized at this stage.

FACET is designed to accept a set of product specifications for manufacturing a family of mechanical gears. This information is accompanied by the batch sizes and production volumes expected for different types of gears within the product family. The specifications in conjunction with a database of production equipment allows FACET to determine the manufacturing operations required for the gears as well as the machines that can implement them. The final step involves a diagram for laying out the machines on the factory floor.

The specifications for an expert system involve not only the functional requirements, but the secondary constraints that define the limits of acceptable solutions. A key constraint for FACET is its compatibility with the computational and memory resources of a personal workstation.

## Space

An expert program is constrained spatially by the computer on which it runs. A simple expert might be self-contained, requiring no interfaces other than the user of the program. Other expert programs may require access to additional processors for heavy-duty number crunching or external sources of data, either within the organization or from off-site utilities.

In addition, an expert system should provide a friendly interface to enable its use by individuals who may be unfamiliar with computerese. Menu-driven interfaces with icons, as well as additional graphics, are often helpful for displaying information in terms familiar to the user.

FACET is designed to respond to commands through a graphic interface and to display information pictorially, including diagrams for the mechanical gears and the factory layout. The system architecture allows for access to information in computer-aided design databases as well as information on production equipment.

## Structure

A modular structure for a knowledge-based system facilitates its construction and enhancement. One way to partition the program is in terms of domain independence. Some knowledge—such as rules for determining gear ratios—relates specifically to the domain of application. Other knowledge is more general, such as the desirability of locating a transfer line near a row of pillars to facilitate the delivery of electricity, communication, and other utilities; this rule of thumb relates not only to gear manufacturing, but to factory design in any industry.

To extend this line of thought further, certain knowledge is entirely independent of the domain. An example is the selection of a highly specific rule over a general one in the interest of enhanced relevance to the problem at hand. Such metalevel rules control the inference procedure to accelerate the reasoning process and even to prevent the program from running into dead ends.

Most programs of significant size evolve over time in response to changing organizational needs. In fact, the total cost of improvements over the lifespan of the evolving program is usually more than three times the figure for the initial working version. Given the rapid pace of technological change in the computer industry, a knowledge system must be designed with an open architecture. The open configuration allows the system to utilize improved capabilities in both hardware and software, whether these are developed within the organization or in the environment at large. The partitioning of the program into comprehensible modules facilitates the process of functional modifications as well as incorporation of novel technologies.

The overall structure of FACET is similar to the generic architecture shown in Figure A.7 of Appendix A. The major distinction is an additional interface in the form of access through a local network to external databases such as equipment specifications.

## Time

A knowledge system must usually respond in realtime to inquiries from a user. On the other hand, it is conceivable that a major computational task, such as the long-run simulation of an entire factory or economy, may proceed offline, in batch mode.

A program that remains immutable will stagnate and lose value over time due to changing environmental circumstances. As discussed previously, a major software package often evolves to adapt to changing user needs. Modification of the program through the addition of capabilities implies an increase in system intelligence.

Even when the repertoire of functions remains static, however, intelligence may be increased through enhanced effectiveness. This may be done in part by learning from previous designs. A straightforward way to implement such changes is for the programmer to update the knowledge base accordingly. On the other hand, promising results in current research toward machine learning indicates that self-programming capabilities will become feasible in the years ahead.

The factory design advisor responds in realtime to most inquiries. The complete design of a complex project, however, may require extensive computation that is better performed as a batch process. The system has been configured in modular fashion to facilitate its expansion over time as well as enhancement through feedback on the performance of previous designs.

## Process

A knowledge system may reason in the forward direction from an initial set of data to determine all its ramifications. An alternate approach is to begin with a hypothesis and reason backward to ascertain whether the data support the propositions. Often, the best strategy is to proceed in both directions until the chains of thought meet at some middle ground.

A complex problem suggests a strategy of divide and conquer. Many problems are amenable to cross-sectional partitioning when the level of interaction among the modules is low. For instance, a computer simulation of a world economy may reason about material flows separately from services or information exchanges. An alternative mode of cross-sectional partitioning in this example is by geography: countries on the Pacific rim might be modeled separately from the European block as well as the Middle East and the southern hemisphere. Eventually these submodels will have to be integrated into a coherent whole.

A second dimension for modularizing a problem is through longitudinal decomposition. A straightforward approach for the simulation example is in terms of economic cycles: recovery followed by expansion, which in turn gives way to recession and contraction. The nature and extent of the partitioning will determine the appropriate level of coordination required to keep track of the computational processes.

The large-scale strategy of FACET is to decompose the work longitudinally into four phases. The first stage involves the *categorization* of the portfolio of gears into groups or types. The gears in each group share characteristics that allow them to be produced on similar manufacturing equipment. The results of this phase may be viewed as a cross-sectional partitioning of the gears into groups, each of which can be addressed separately.

The second phase involves *process planning*. This is a logical identification of the production operations required to manufacture the product.

The next phase involves a *time study* of the production process. This may be viewed as a physical process plan to determine the time required for each operation, information that is instrumental in determining the capacity and number of machines required in a production line. For instance, if Step $B$ is required of all gears passing through Step $A$, but requires twice as much time as the earlier process, then the throughput capacity of the equipment must be twice as high; otherwise bottlenecks will occur.

The final phase of the advisor is the *layout* of the machines on the factory floor. These machines may be arranged into one or more production lines, depending on the requirements of the particular factory.

### *Efficiency*

A knowledge system should be efficient in processing as well as memory requirements. Processing time is kept in check through domain-dependent knowledge. Such knowlege results in the reduction of a potentially enormous number of alternative paths into one of manageable size.

Some techniqes reduce computational requirements without loss of optimality. A well-known example is the binary search of an ordered list, in which the space of alternatives is reduced by half at each stage. For this technique, the number of inferences required grows only logarithmically as the size of the table expands.

On the other hand, much domain knowledge such as the guidelines used by human experts in investment finance, do not guarantee an optimal solution. These rules of thumb specify, instead, how a solution of acceptable utility can be obtained at reasonable processing cost.

Some programs rely on extensive knowledge bases for their proper operation. This problem can be ameliorated by using compact data structures. However, the compaction of data may result in slowing down the inference procedures if the information cannot be readily interpreted. One way to circumvent these difficulties is to maintain information externally, on neighboring computers linked by a local network or even external utilities accessed by a global network. The most effective strategy will be determined by the specific application.

The factory planning advisor minimizes computational requirements through special formatting. Salient aspects of the most heavily-utilized items of knowledge are integrated by a pre-processing module into a tree structure that can be rapidly accessed by the inference modules. An example of a popular item of knowledge is the set of candidate machines and the manufacturing operations they can effect; formatting this information into an ordered tree can facilitate the search for particular items of data.

In addition, the expert system minimizes the requirements on main memory through interfaces with external sources. For instance, the complete specifications for the family of gears can be obtained by accessing a computer-aided design base through a local network.

## Summary

Artifacts are conceived, designed, and produced by intelligent systems. The basic object may itself be intelligent, as in a robot, or mindless, as in a bridge. The design of any product implies a set of levels of intelligence. The plant that constructs the product is a smart system that was in turn produced by a second-level intelligence. If the second-level intelligence is a program rather than a person, it too was developed by a higher-order entity.

There are current efforts to develop a third-order program that incorporates strategies and tools to build expert software. At first these programs will merely assist human designers, but in the long run they will likely incorporate learning capabilities and become capable of autonomous operation.

The future will reveal how far these layers of intelligent design can usefully go. Whatever the final destination, however, the factors of intelligence discussed in this book can serve as a vehicle.

# Notes

References are denoted by parentheses. For example Kim (1985) or (Kim 1985) denotes the publication that appeared in 1985 and is listed in the Selected Bibliography of this book. When two or more publications are by the same author in the same year, each is distinguished by an alphabetic suffix after the year, as in Kim (1987a) and Kim (1987b).

## Part I

1. In Decartes (1973), p. 2. The *Regulae Ad Directionem Ingenii* may have been written in the winter of 1628–9. However, this is an unfinished piece whose precise date of composition is unknown. The original manuscript is lost, and published accounts are based on secondary sources.

## Chapter 1

1. From a manuscript entitled "Scheme for Establishing the Royal Society," quoted in David Brewster, *Memoirs of the Life, Writings, and Discoveries of Sir Isaac Newton*, 2 vols., Edinburgh: Constable and Co., 1855, vol. I, p. 102.
2. See Kim (1987b).
3. Kim (1988e).
4. Ashby (1956), Ch. 6.
5. Minsky (1986).
6. Luria (1975), p. 359.
7. Simon (1985), p. 36.
8. Darwin (1964), p. 206.
9. Moravec (1988), p. 167.
10. Dawkins (1976), p. 2.

## Chapter 2

1. *On the Origin of Species* (1959), Ch. XIV; in Darwin (1964), p. 490.
2. A collection of random events is perceived as noise. Examples of random signals are static on the radio or "snow" on television.
3. For the mathematically inclined reader, the relationships among the attributes of the general framework may be given a geometric interpretation. In the language of linear algebra,

the attributes may be viewed as a set of vectors that define the state space of the design situation. These vectors form a basis or minimal spanning set which together characterize the critical aspects of the problem. The set of vectors is noncolinear, since no single vector can be described as some linear combination of the others. On the other hand, the vectors tend to be nonorthogonal to each other, since the choice of an operating point on one vector can affect the optimal value on another.

## Part II

  1. Luria (1975), p. 308.

## Chapter 3

1. *On the Origin of Species* (1859), Ch. VI; in Darwin (1964), p. 200.
2. Ackoff and Emery (1972), p. 31.
3. Kim (1985).

## Chapter 4

1. Drucker (1980), p. 1.
2. Ashby (1956), p. 207.
3. The separation of components into apparent and hidden facets is also referred to as the black box and gray box models, respectively.
4. Emery and Trist (1965)
5. Summers (1980).
6. Lawrence and Lorsch (1967).
7. Cyert and March (1963), Ch. 6.
8. As with many classifications for intelligent systems, the distinction between object level and metalevel are artifical rather than real. They are convenient labels for discussion; in reality they represent merely two points on a smooth scale.
9. Winchester and Mertens (1983), p. 254.
10. Pascale and Athos (1981), pp. 43–47.
11. Pascale and Athos (1981), p. 47.
12. Simon (1985), pp. 63–64.
13. Braitenberg (1984).
14. Braitenberg (1984), pp. 6–9 and 26–28.

## Chapter 5

1. *Psychology: Briefer Course* (1892), Ch. VIII; as given in James (1984), p. 101.
2. Minsky (1986), p. 17.
3. Wiener (1967), pp. 79–80.
4. Miller (1978), pp. 317–319.
5. Miller (1978), pp. 285–286; Lehninger (1975), pp. 663–65; Marten *et al.* (1985), pp. 214–217.
6. Thompson (1984), pp. 234–236.

7. Grundfest (1940); Groves and Schlesinger (1979), p. 46; Vander *et al.* (1985), p. 206.

8. Grundfest (1940); Groves and Schlesinger (1979), p. 46; Vander *et al.* (1985), p. 206.

9. Vander *et al.* (1985), pp. 205–206.

10. In the field of graph theory, a network is called a *graph*. A node is also called a *vertex*; a link is also known as an *arc* or an *edge*. The algebraic representation is as follows. Let $N$ be a set of nodes $\{n_1, n_2, \ldots\}$. An edge is any pair $\langle n_i, n_j \rangle$ of nodes from $N$; now let $E$ be the set of such edges. Then the ordered pair $\langle N, E \rangle$ defines a graph $G$. Further, a directed edge representing a unilateral link is denoted by the ordered pair $\langle n_i, n_j \rangle$. A graph that contains directed edges is called a directed graph, or simply *digraph*.

11. Quillian (1968); Rich (1983), p. 215.

12. Kim (1987d).

13. Vander *et al.* (1985), p. 177; Angevine and Cotman (1981), pp. 217–219.

14. Miller (1978), pp. 392–393.

15. Noback and Demarest (1981), pp. 37; 81–82.

16. Lindsley and Holmes (1984), pp. 101–102.

17. Hillis (1985), p. 4.

18. Hillis (1985).

19. Clocksin and Mellish (1984).

20. Chandler (1962).

## Chapter 6

1. Prigogine (1980), p. xvii.

2. Asimov (1952), p. 7.

3. Bratko (1986); Clocksin and Mellish (1984); Kowalski (1974, 1979).

4. Schrödinger (1967), p. 79.

5. Bohm and Peat (1987), pp. 130–131.

6. The oscillation period is on the order of one minute.

7. Prigogine (1980), pp. 120–122.

8. Drucker (1981), p. 47.

9. Toynbee (1947), p. 555.

10. Toynbee (1947), p. 246.

11. Churchland (1984), p. 153.

12. Bohm and Peat (1987), p. 134.

13. Vander *et al.* (1985), p. 211.

14. Vander *et al.* (1985), pp. 62; 69–70.

15. Vander *et al.* (1985), p. 69.

16. Galbraith (1973), Ch. 3.

17. Cyert and March (1963), Ch. 6.

18. Kim (1985); Lawson (1981).

19. Vander *et al.* (1985), p. 635.

20. Vander *et al.* (1985), pp. 99, 282–285.

## Chapter 7

1. Barnard (1938), p. 6.

2. Russell (1927), p. 292.

3. Russell (1927), p. 287.

4. Kandel and Schwartz (1985), pp. 837, 841.

5. Vander *et al.* (1985), pp. 483–485.

6. Vander *et al.* (1985), pp. 652–653.

7. Clocksin and Mellish (1984); Kowalski (1979).

8. McClelland *et al.* (1986), pp. 347–348; Angevine and Cotman (1981), pp.166–167, 176–177, 180–181.

9. Cortical outputs do not have to rely on thalamic neurons. The axons projecting from the cortex can lead directly to the spinal cord, for example, whereas the inputs from the spinal cord to the cortex must relay through the thalamus.

10. Shepherd (1983), p. 117.

11. Therman (1986), p. 262; Miller (1978), p. 351.

12. Angevine and Cotman (1981), pp. 81–83.

13. Barnard (1938), p. 239.

14. Chandler (1962), pp. 78–89.

15. Chandler (1962), p. 104.

16. Chandler (1962), p. 129.

17. Chandler (1962), p. 142.

18. Arrow (1951), p. 59.

## Chapter 8

1. *On the Origins of Species* (1959), Ch. VI; in Darwin (1964), p. 206.

2. von Mises (1962), pp. 7–8.

3. A number of important issues have eluded attempts toward monetarization. The gross national product is a measure of the total value of all the goods and services produced in an economy. But what are the true costs of producing this value? Some costs are difficult to quantify, such as the detrimental consequences of pollution or the welfare of a society whose members enjoy robust health. The net economic welfare has been proposed as a more useful measure of the value of production, but it has its own drawbacks. Issues such as these provide fertile ground for debate among economists.

4. "Man is subject to the passing of time. . . . He must economize it as he does other scarce factors. The economization of time has a peculiar character because of the uniqueness and irreversibility of the temporal order." von Mises (1949), p. 101.

5. Drexler (1986).

6. Luciano *et al.* (1983), pp. 452–454; Angevine and Cotman (1981), p. 357.

7. Schmidt (1985), pp. 103–104.

## Part III

1. Smith (1802), vol. II, p. 59.

2. von Neumann (1966), p. 63.

## Chapter 9

1. Einstein (1945), pp. 30–31.

2. Lindsley and Holmes (1984), p. 178; Luciano *et al.* (1983), p. 343.

## Chapter 10

1. *Psychology: Briefer Course* (1892), Ch. I; in James (1984), p. 11.
2. Wiener (1967), pp. 79–80. Italics in original.
3. Luciano *et al.* (1983), p. 612.
4. Albus (1981), p. 30.

## Part IV

1. Peters (1987), p. 36.

## Chapter 11

1. Asimov (1976), p. 144.
2. Capek (1923).
3. Asimov (1950).
4. Smith and Sciaky (1988).
5. Nilsson and Rosen (1969).
6. Moravec (1981).
7. Mosher (1969).
8. Raibert et al. (1984a, 1984b).
9. Raibert (1986).
10. The human nervous system contains perhaps 100 billion neurons, each of which may have a thousand synapses on average. If each synapse can take any one of a thousand states, it can encode the equivalent of 10 binary digits. The resulting memory capacity is $10^{15}$ bits.

What about the processing power? The assemblage of $10^{11}$ neurons can change state in 10 milliseconds, or about 100 times per second. The total processing power is then $10^{13}$ operations per second. This figure can be multiplied by the thousand synapses corresponding to 10 bits. For each neuron, the outgoing signals to the synapses will be identical; but the incoming signals at the dendrites will be highly variant. The resulting computational power of the nervous system is $10^{14}$ bits per second.

These figures represent only gross estimates, but they serve our purpose even if they are off-target by an order of magnitude or two. On the other hand, the numbers are comparable to those obtained through an analysis of visual information processing, as discussed by Moravec (1988), pp. 59–61.
11. Brooks (1987).
12. 1 trillionth of a second.

## Chapter 12

1. Smith (1802), vol. I, p. 2.
2. von Neumann (1966), p. 80.
3. Maley (1988).
4. These figures relate to competent, professional programmers. A few exceptional programmers can produce over 10 times more results than those of average ability.
5. In comparison, the operating system for the IBM 360 family of computers required

approximately 5000 work years for its design, coding and documentation from 1963 to 1966 (Brooks 1975), p.31.

6. Pao and Jelinek (1988); *Business Week*, Nov. 13, 1989, p. 114.


## Chapter 13

1. Sculley (1987), p. 420.
2. Sculley (1987), p. 265.
3. Sculley (1987), p. 388.
4. Butcher (1988), p. 190.
5. Butcher (1988), p.211.


## Part V

1. Russell (1929), p. 312.


## Chapter 14

1. Plato, *Republic*, Bk. IV, sec. 441.

2. "If the possession of information can be understood as the possession of some internal physical order that bears some systematic relation to the environment, then the operations of intelligence, abstractly conceived, turn out to be just a high-grade version of the operations characteristic of life, save that they are even more intricately coupled to the environment." Churchland (1984), p. 154.

3. Clarke (1986), p. 71.

4. "Our genes may be immortal but the *collection* of genes which is any one of us is bound to crumble away. Elizabeth II is a direct descendant of William the Conqueror. Yet it is quite probable that the she bears not a single one of the old king's genes. We should not seek immortality in reproduction. But if you contribute to the world's culture, if you have a good idea, compose a tune, invent a sparking plug, write a poem, it may live on, intact, long after your genes have dissolved in the common pool." Dawkins (1976), p. 214. Italics in original.

5. Drexler (1986), pp. 231–232.

6. Kennedy, in an address at the University of California, Berkeley, 23 March 1962; as given in Beck (1980), p. 891.


## Appendices

1. Churchland (1984), p. 154.
2. Dawkins (1976), p. 214. Italics in original.


## Appendix A

1. Simon (1985), p. 159.
2. Kim (1988c).

3. Lenat (1983).
4. Lenat (1982).
5. Kant and Newell (1984); Mitchell *et al.* (1981); Rich *et al.* (1979).
6. Balzer *et al.* (1976); Mostow (1983, 1985); Scherlis and Scott (1983)
7. Bittinger (1982); Hamilton (1978); Kneebone (1963).
8. Glorioso and Osorio (1980); von Neumann (1966).
9. Hopcroft and Ullman (1979); Minsky (1967).
10. Adapted from Parthasarathy and Kim (1987).
11. Kim (1985, 1986b).
12. Kim (1985); Kim and Suh (1987); Parthasarathy and Kim (1987).
13. Ulrich and Seering (1988).
14. Lenat (1982, 1983).
15. Kim (1985); Parthasarathy and Kim (1987).
16. Kim (1987b); Kim, Hom, and Parthasarathy (1988).
17. Kim (1985); Kim and Suh (1987); Kim, Hom, and Parthasarathy (1988).

## Appendix B

1. Whitehead (1938), p. 42.
2. Kim (1985).
3. Bittinger (1982); Hamilton (1978); Kneebone (1963); Kowalski (1979).
4. Kim (1985).
5. Kim (1989b).
6. Compare Genesereth and Nilsson (1987).

## Appendix C

1. *Novum Organum* (1620), Bk. I, sec. xix; as given in Bacon (1905), p. 261. Italics added.
2. Kim (1985); Kim and Suh (1985, 1987).
3. Suh (1985); Suh et al. (1978).
4. Suh (1984, 1990).
5. Kim (1985); Kim and Suh (1985, 1987).
6. Suh (1985, 1990).
7. Suh et al. (1978a, 1978b).
8. Kim and Suh (1989).
9. Kim (1985); Kim and Suh (1985).
10. Clocksin and Mellish (1981); Gallaire and Minker (1978); Kowalski (1974); Pereira (1984).
11. Horstmann (1983); Mizoguchi (1983); Oliveira (1984); Walker (1983).
12. Kim (1985, 1987b); Kim et al. (1985, 1987, 1988).
13. Newell and Simon (1972).
14. Kim (1985); Kim and Suh (1985, 1987).

## Appendix D

1. *Analects*, Bk. XVIII, sec. 8; as given in Confucius (1938), pp. 211–212.
2. Holland (1975).

3. This appendix is based on Kim (1988d).

4. Kim (1988d).

5. Deming (1982); etc.

6. Kim (1985).

7. Glorioso & Osorio (1980).

8. Kim (1985); Parthasarathy & Kim (1987).

9. Holland (1975).

10. Holland and Reitman (1978).

11. Goldberg (1983).

12. Pao (1988).

13. In this vein, one promising approach may be the iterative generation of aggregated concepts as implemented in the BACON program (Langley *et al.* 1983). This program seeks patterns among sets of data pertaining to different parameters or variables. The program has prior knowledge of the candidate algebraic relationships, such as the multiplication or division of variables, and seeks out the actual relationships consistent with the data sets. This technique has been used to "rediscover" a number of algebraic relationships in chemistry and physics, such as Ohm's Law which specifies voltage as the product of current and resistance.

## Appendix E

1. Einstein (1938), p. 291.

2. Kim (1985, 1987c, 1987d, 1988a).

3. A more systematic derivation of the logarithmic relationship between $I_i$ and the reciprocal of $p_i$ can be given. For example, see Appendix D of Kim (1985).

4. Machlup and Mansfield (1983); Questler (1955).

5. Kim (1985, 1987); Nakazawa and Suh (1984).

6. Galbraith (1973).

7. Tsypkin (1971).

8. Kuo (1980), p. 694.

## Appendix F

1. Whitehead (1938), pp. 214–215.

2. Kim and Tokawa (1989); Tokawa (1989).

# Selected Bibliography

Ackoff, R. L. "Towards a Behavioral Theory of Communication." In *Management Science*, v.4, 1957–58: 218–34.

Ackoff, R. L. and R. L. Emery. *On Purposeful Systems*. London: Travistock,1972.

Agha, G. *Actors: A Model of Concurrent Computation in Distributed Systems*. Cambridge, MA: MIT Press, 1987.

Albus, J. S. *Brains, Behavior and Robotics*. Peterborough, NH: Byte, 1981.

Alexander, Igor. *Designing Intelligent Systems*. NY: UNIPUB, 1984.

Allison, G. T. *Essence of Decision*. Boston: Little Brown, 1971.

Anderson, R. L. *A Robot Ping-Pong Player*. Cambridge, MA: MIT Press, 1986.

Angevine, J. B., Jr., and C. W. Cotman. *Principles of Neuroanatomy*. Oxford: Oxford Univ. Press, 1981.

Ansoff, H. I. *Strategic Management*. NY: Wiley, 1979.

Arrow, K. J. *Social Choice and Individual Values*. NY: Wiley, 1951.

Ashby, W. R. *An Introduction to Cybernetics*. NY: Wiley, 1956.

Ashby, W. R. *Design for a Brain*. London: Chapman and Hall, 2nd ed., 1960, rpt. 1978.

Asimov, I. *I, Robot*. Garden City, NY: Doubleday, 1950.

Asimov, I. *Foundation and Empire*. NY: Avon, 1952.

Asimov, I. *The Bicentennial Man . . . and Other Stories*. NY: Doubleday, 1976.

Bacon, F. *The Philosophical Works of Francis Bacon*. Ed. by J. M. Robertson. London: Routledge, 1905.

Baker-Ward, L., M. H. Feinstein, J. L. Garfield, E. L. Rissland, D. A. Rosenbaum, N. A. Stillings, and S. E. Weisler. *Cognitive Science: An Introduction*. Cambridge, MA: MIT Press, 1987.

Balzer, R., N. Goldman, and D. Wile. "On the Transformational Implementation Approach to Programming." *Proceedings of the Second International Conference on Software Engineering*, 1976: 337–343.

Barnard, C. I. *The Functions of the Executive*. Cambridge, MA: Harvard Univ. Press, 1938.

Battani, G. and H. Meloni. "Interpreteur du Language de Programation PROLOG." Groupe de l'Intelligence Artificielle, U.E.R. de Luminy, Marseille, France, 1973.

Beer, Stafford. *Decision and Control*. NY: Wiley, 1966.

Beer, Stafford. *Cybernetics and Management*, 2nd ed. London: English Universities Press, 1967.

Beer, Stafford. *Brain of the Firm: The Managerial Cybernetics of Organization*. London: Penguin, 1972.

Bittinger, M. *Logic, Proof and Sets*. Reading, MA: Addison-Wesley, 1982.

Bjorke, O. "Manufacturing System Theory—What is That?" *Annals of the CIRP*, v. 22 (1), 1973: 157–158.

Bjorke, O. "Towards Integrated Manufacturing Systems—Manufacturing Cells and their Subsystems." *Robotics and Computer-Integrated Manufacturing*, v. 1(1), 1984: 3–20.

Bjorke, O. and O. I. Franksen. "Manufacturing Systems Theory and its Relations to the Basic Disciplines of Science—From Measurement to Systems." SINTEF Report, Trondheim, Norway, 1985.

Bohm, David and F. David Peat. *Science, Order and Creativity*. NY: Bantam, 1987.

Border, J. and N. P. Suh. "Intelligent Injection Molding." Society of Plastics Engineers, *ANTEC '82*, v. 28, 1982: 323–326.

Boulding, K. "General Systems Theory—the Skeleton of Science." *Management Science,* v.2, 1956: 197–208.

Brady, J. M. *The Theory of Computer Science—A Programming Approach*. London: Chapman and Hall, 1977.

Braitenberg, V. *Vehicles: Experiments in Synthetic Psychology*. Cambridge, MA: MIT Press, 1984.

Bratko, I. *PROLOG Programming for Artificial Intelligence*. Reading, MA: Addison-Wesley, 1986.

Brillouin, L. *Science and Information Theory*. NY: Academic, 1960.

Brooks, F. P. Jr. *Mythical Man-Month*. Reading, MA: Addison-Wesley, 1975.

Brooks, R. A. "A Robust Layered Control System for a Mobile Robot." *Journal of Robotics and Automation*, v. RA-2 (1), March 1986: 14–23.

Brooks, R. A. "Autonomous Mobile Robots." In *AI in the 1980's and Beyond: An MIT Survey*. Cambridge, MA: MIT Press, 1987.

Buckley, W., ed. *Modern Systems Research for the Behavioral Scientist*. Chicago: Aldine, 1968.

Bullock, T. H. "Signals and Neuronal Coding." In *The Neurosciences: A Study Program*. Edited by Gardner C. Quarton, Theodore Melnechuk, and Francis O. Schmitt. New York: Rockefeller Univ. Press, 1967: 347–51.

Bunge, M. *Ontology I: The Furniture of the World*. Boston: Reidel, 1977.

Bunge, M. *Ontology II: A World of Systems*. Boston: Reidel, 1979.

Burks, A. W. *Essays on Cellular Automata*. Urbana: University of Illinois, 1970.

Butcher, L. *Accidental Millionaire: The Rise and Fall of Steve Jobs at Apple Computer*. NY: Paragon, 1988.

Capek, K. *R.U.R.* (Rossum's Universal Robots). London: Samuel French, 1923.

Chandler, A. D. *Strategy and Structure*: *Chapters in the History of the Industrial Enterprise*. Cambridge: MIT Press, 1962.

Churchland, P. M. *Matter and Consciousness*. Cambridge, MA: MIT Press, 1984.

Churchland, P. S. *Neurophilosophy: Toward a Unified Science of the Mind/Brain*. Cambridge, MA: MIT Press, 1986.

Clarke, A. C. *July 20, 2019: Life in the 21st Century*. NY: Macmillan, 1986.

Clocksin, W. F., and C. S. Mellish. *Programming in Prolog,* 2nd ed. New York: Springer-Verlag, 1984.

Confucius. *The Analects of Confucius*. Trans. by A. Waley. NY: Vintage, 1938.

Cothier, P. H. "Assessment of Timeliness in Command and Control." Laboratory for Information and Decision Sciences, Report LIDS-TH-1391, M.I.T., Aug. 1984.

Courtois, P. J. *Decomposability: Queuing and Computer System Applications*. NY: Academic, 1977.

Cyert, R. and J. March. *A Behavioral Theory of the Firm*. Englewood Cliffs, NJ: Prentice-Hall, 1963.

Darwin, C. R. *On the Origin of Species*. Cambridge, MA: Harvard Univ. Press, 1964.

Davis, M. *Computability and Unsolvability*. NY: McGraw-HIll, 1958.

Dawkins, Richard. *The Selfish Gene*. Oxford: Oxford Univ. Press, 1976.

Deming, W. E. *Quality, Productivity and Competitive Position*. Cambridge, MA: M.I.T. Center for Advanced Engineering Studies, 1982.

Dennett, Daniel C. *Brainstorms*. Cambridge, MA: MIT Press, 1988.

Descartes, R. "Rules for the Direction of the Mind" (*Regulae Ad Directionem Ingenii*). In E. Haldone and G.R.T. Ross, translators, *The Philosophical Works of Descartes*, v.I, 1911; rpt. Cambridge, UK: Cambridge Univ. Press, 1973.

Dorf, R. C. *Modern Control Systems*, 2nd ed. Reading, MA: Addison-Wesley, 1974.

Drexler, K. E. *Engines of Creation*. NY: Doubleday, 1986.

Drucker, P. F. *Managing in Turbulent Times*. NY: Harper and Row, 1980.

Drucker, P. F. *Toward the Next Economics and Other Essays*. NY: Harper and Row, 1981.

Egilmez, K., and S. H. Kim. "Design and Control of Manufacturing Systems: An Integrated Approach Based on Model Logic and Zonecharts." *Robotics and Computer-Integrated Manufacturing*, v.6(3), 1989: 209–228.

Einstein, A. *The Meaning of Relativity*. Princeton: Princeton University Press, 1945.

Einstein, A., with L. Infeld. *The Evolution of Physics*. NY: Simon and Schuster, 1938.

Emery, F. E. and E. L. Trist. "The Causal Texture of Organizational Environments." In *Human Relations*, v.18, Feb. 1965: 21–31. Reprinted in M.B. Brinkerhoff and P.R. Kunz, *Complex Organizations and their Environments*, Dubuque, IA: W.C. Brown, 1972.

Everett, H. R. "Security and Sentry Robots." In R. C. Dorf, ed., *International Encyclopedia of Robotics*. NY: Wiley, 1988: 1462–1476.

Fields, W. S., and W. Abbott. *Information Storage and Neural Control*. Springfield, IL: Charles C. Thomas, 1963.

Firschein, O., and M. A. Fischler. *Intelligence: The Eye, the Brain, and the Computer*. Reading, MA: Addison-Wesley, 1987.

Fisher, R. A. *The Genetical Theory of Natural Selection*, 2nd ed. NY: Dover, 1958.

Forsyth, R., ed. *Expert Systems: Principles and Case Studies*. New York: Chapman and Hall, 1984.

Fortini, E. T. *Dimensioning for Interchangeable Manufacture*. NY: Industrial Press, 1967: 51.

Freeman, P. and A. Newell. "A Model for Full Reasoning in Design." *Proc. Second International Joint Conference on Artificial Intelligence*, London: Sept. 1971: 621–633.

Galbraith, J. *Designing Complex Organizations*. Reading, MA: Addison-Wesley, 1973.

Gallagher, R. E. *Information Theory and Reliable Communications*. NY: Wiley, 1968.

Gallaire, H. and J. Minker, eds. *Logic and Databases*. NY: Plenum, 1978.

Gardner, H. *Frames of Mind*. NY: Basic, 1983.

Gefenstette, J. J. *Proc. International Conference on Genetic Algorithms and their Applications*. Carnegie-Mellon University, Pittsburgh, PA: July,1985.

Genesereth, M. R. and N. J. Nilsson. *Logical Foundations of Artificial Intelligence*. Los Altos, CA: M. Kaufmann, 1987.

Georgescu-Roegen, N. *The Entropy Law and the Economic Process*. Cambridge, MA: Harvard Univ. Press, 1971.

Gheorge, A. V. *Applied Systems Engineering*. NY: Wiley, 1981.

Ginzberg, A. *Algebraic Theory of Automata*. NY: Academic, 1968.

Glorioso, R. M. and F. C. Colon Osorio. *Engineering Intelligent Systems: Concepts, Theory and Applications*. Bedford, MA: Digital Press, 1980.

Glueck, J. *Chaos: Making a New Science*. NY: Viking, 1987.

Glueck, W. F. *Business Policy and Strategic Management*, 3rd ed. NY: McGraw-Hill, 1980.

Goldberg, D. E. "Computer-Aided Gas Pipeline Operation Using Genetic Algorithms and Rule Learning." Ph.D. thesis, University of Michigan, Ann Arbor, MI: 1983.

Groover, M. P. *Automation, Production Systems, and Computer-Aided Manufacturing*. Englewood Cliffs, NJ: Prentice-Hall, 1980.

Groover, M. P. and E. W. Zimmers, Jr. *CAD/CAM: Computer-Aided Design and Manufacturing*. Englewood Cliffs, NJ: Prentice-Hall, 1984: Chapter 20.

Groves, P. and K. Schlesinger. *Introduction to Biological Psychology*. Dubuque, IA: Brown, 1979.

Grundfest, H. "Bioelectric Potentials." In *Annual Review of Physiology, v.2.* Ed. by J.M. Luck, Stanford: American Physiological Society and Annual Reviews, Inc., 1940, pp. 213–242.

Guiasu, S. *Information Theory with Applications*. NY: McGraw-HIll, 1977.

Hagler, C. D. and S. H. Kim, "Information and its Effect on the Performance of a Robotic Assembly Process," *Proc. Symp. on Intelligent and Integrated Manufacturing: Analysis and Synthesis,* Dec. 1987: 349–356.

Haken, H. *Synergetics*. NY: Springer-Verlag, 1978.

Halberstam, David. *The Reckoning*. NY: Avon, 1986.

Hamilton, A. G. *Logic for Mathematicians*. Cambridge, U.K.: Cambridge Univ. Press, 1978.

Hardt, D. E. "Shape Control in Metal Bending Processes: the Model Measurement Tradeoff." In D. E. Hardt, ed., *Information Control Problems in Manufacturing Technology,* Proc. Fourth IFAC Symposium, 1982.

Harmon, P. and D. King. *Expert Systems: Artificial Intelligence in Business*. NY: Wiley, 1985.

Hax, A. C. and N. S. Majluf. "The Corporate Strategic Planning Process." Sloan School of Management, W.P. 1396–83, M.I.T., Cambridge, MA, 1983.

Hax, A. C. and N. S. Majluf. *Strategic Management: An Integrative Perspective*. Englewood Cliffs, NJ: Prentice-Hall, 1984.

Hillis, W. D. *The Connection Machine*. Cambridge, MA: MIT Press, 1985.

Hiltz, S. R. *Online Communities: A Case Study of the Office of the Future*. Norwood, NJ: Ablex, 1984.

Hiltz, S. R. and M. Turoff. *The Network Nation*. Reading, MA: Addison-Wesley, 1978.

Holland, J. H. *Adaptation in Natural and Artificial Systems*. Ann Arbor: University of Michigan Press, 1975.

Holland, J. H. and J. S. Reitman. "Cognitive Systems Based on Adaptive Algorithms." In D. A. Waterman and F. Hayes-Roth, eds., *Pattern-Directed Inference Systems,* New York: Academic, 1978: 313–329.

Holland, J. H. and J. S. Reitman. "Escaping Brittleness: the Possibilities of General-Purpose Learning Algorithms Applied to Rule-Based Systems." In R. S. Michalski et al., eds., *Machine Learning: An Artificial Intelligence Approach, v. II,* Los Altos, CA: M. Kaufmann, 1986: 593–624.

Hopcroft, J. E. and J. D. Ullman. *Introduction to Automata Theory, Languages, and Computation*. Reading, MA: Addison-Wesley, 1979.

Horstmann, P. W. "Expert Systems and Logic Programming for CAD. " *VLSI Design*, Nov. 1983: 37–46.

House, E. L., B. Pansky, and A. Siegel. *A Systematic Approach to Neuroscience*. New York: McGraw-Hill, 1979.

Hsu, J. C. and A. U. Meyer. *Modern Control Principles and Applications*. NY: McGraw-Hill, 1968.

Jackson, P. *Introduction to Expert Systems*. Reading, MA: Addison-Wesley, 1986.

Jacobson, G. and J. Hillkirk. *Xerox: American Samurai*. NY: Macmillan, 1986.

James, W. *Psychology: Briefer Course*. Cambridge, MA: Harvard Univ. Press, 1984.

Kandel, E. R. and J. H. Schwartz, eds. *Principles of Neural Science*, 2nd ed. New York: Elsevier, 1985.

Kant, E. and A. Newell. "Problem Solving Techniques for the Design of Algorithms." Technical Report CMU-CS-82-145, Computer Science Department, Carnegie-Mellon University. Also in *Information Processing and Management*, London: Pergamon Press, 1984.

Kanter, J. *Management Information Systems*, 3rd ed. Englewood Cliffs, NJ: Prentice-Hall, 1984.

Kim, S. H. "Mathematical Foundations of Manufacturing Science: Theory and Implications." Ph.D. Thesis, Cambridge, MA: MIT, May 1985.

Kim, S. H. "Frameworks for a Science of Manufacturing." *Proc. North American Manufacuring Research Conference*, Minneapolis, MN: May, 1986a: 552–557.

Kim, S. H. "A Mathematical Framework for Intelligent Manufacturing Systems." *Proc. Symposium on Integrated and Intelligent Manufacturing*. ASME Winter Annual Meeting, Anaheim, CA: Dec., 1986b: 1–8.

Kim, S. H. "A Framework for Research in Manufacturing Systems." Technical Report, Laboratory for Manufacturing and Productivitiy, MIT, Cambridge, MA: 1987a.

Kim, S. H. "A Unified Architecture for Design and Manufacturing Integration." Technical Report, Laboratory for Manufacturing and Productivity, MIT, Cambridge, MA: 1987b.

Kim, S. H. "Managing Complexity in Automated Systems: an Information Measure and its Applications." *Manufacturing Technology Review and North American Manufacturing Research Conference Proc.*, Bethlehem, PA: May, 1987c: 541–545.

Kim, S. H. "A Generalized Information-Theoretic Framework for Intelligent Systems." Technical Report, Laboratory for Manufacturing and Productivitiy, MIT, Cambridge, MA: 1987d.

Kim, S. II. "Information Framework for Robot Design." In R. C. Dorf, ed., *International Encyclopedia of Robotics*. NY: Wiley, 1988a: 653–663.

Kim, S. H. "An Automata-Theoretic Framework for Intelligent Systems." *Robotics and Computer-Integrated Manufacturing*. v.5(1), 1988b: pp. 43–51.

Kim, S. H. "A General Model of Design: Formalization and Consequences." *Manufacturing Review, v. 1(2)*, June 1988c: 109–116.

Kim, S. H. "A Unified Framework for Self-Learning Systems." *Proc. Manufacturing International '88, v. III: Symp. on Manufacturing Systems*. Atlanta, GA April 1988d: 165–170.

Kim, S. H. "Toward a Science of Intelligent Systems". *Advances in Manufacturing System Engineering—1988*, ASME Winter Annual Meeting, Chicago, IL, Nov. 1988e: 27–33.

Kim, S. H. "A Systematic Approach to Intelligent System Design; Part 1: Concepts." *Robotics and Computer-Integrated Manufacturing*, v. 6(2), 1989a: 143–155.

Kim, S. H. "Openness of Reasoning in Intelligent Systems: Logical Closure vs. Domain Dependence." *International J. of Computer Applications in Technology*, v. 2(4), 1989b: 228–233.

Kim, S. H., S. Hom and S. Parthasarathy. "Design and Manufacturing Advisor for Turbine Disks." *Robotics and Computer-Integrated Manufacturing*, v. 4 (3/4), 1988: 585–592.

Kim, S. H. and N. P. Suh. "Application of Symbolic Logic to the Design Axioms." *Robotics and Computer-Integrated Manufacturing*, v. 2(1), 1985: 55–64.

Kim, S. H. and N. P. Suh. "Mathematical Foundations for Manufacturing." *Journal of Engineering for Industry*, Trans. ASME, v.109(3), 1987: 213–218.

Kim, S. H. and N. P. Suh. "Formalizing Decision Rules for Engineering Design." In A. Kusiak, ed., London: Taylor and Francis, 1989: 33–44.

Kim, S. H. and T. Tokawa. "A Systematic Approach to Intelligent System Design; Part 2: Multi-level Case Study." *Robotics and Computer-Integrated Manufacturing*, v. 6(2), 1989: 157–165.

Kneebone, G. T. *Mathematical Logic and the Foundations of Mathematics*. NY: Van Nostrand, 1963.

Koch, C., T. Poggio, and V. Torre. "Computations in the Vertebrate Retina: Gain Enhancement, Differentiation and Motion Discrimination," *Trends in Neurosciences*, v. 9, 1986: 204–11.

Kohonen, T. *Self-Organization and Associative Memory*. New York: Springer-Verlag, 1984.

Kompass, E. J. and T. J. Williams, eds. *Learning Systems and Pattern Recognition*. Barrington, IL: Technical Pub., 1983.

Kowalski, R. "Predicate Logic as a Programming Language." DCL Memo 70, Edinburgh University, Edinburgh, UK, 1974.

Kowalski, R. *Logic for Problem Solving*. NY: North Holland, 1979.

Kuhn, T. S. *The Structure of Scientific Revolutions*. Chicago: University of Chicago, 1970.

Kuo, B. C. *Digital Control Systems*. NY: Holt, Rinehart and Winston, 1980.

Kutcher, M. "Automating it All." *Spectrum*, v.20(5), May 1983: 40–43.

Langley, P., G. L. Bradshaw, and H. A. Simon. "Rediscovering Chemistry with the BACON System." In R. S. Michalski, J. G. Carbonell and T. M. Mitchell, *Machine Learning: An Artificial Intelligence Approach*, Palo Alto, CA: Tioga, 1983: 307–329.

Lawrence, P. and J. Lorsch. *Organizations and Environment: Managing Differentiation and Integration*. Boston: School of Bus. Admin., Harvard Univ., 1967.

Lawson, J. S., Jr. "The Role of Time in a Command Control System." In M. Athans et al., *Proc. Fourth MIT/ONR Workshop on Distributed Information and Decision Systems Motivated by Command-Control-Comunications Problems*, v. 4, Laboratory for Information and Decision Systems, Report LIDS-R-1159, MIT, Oct., 1981: 19–60.

Lehninger, A. L. *Biochemistry*, 2nd ed. NY: Worth, 1975.

Lenat, D. "AM: Discovery in Mathematics as Heuristic Search." In R. Davis and D. B. Lenat, eds., *Knowledge-Based Systems in Artificial Intelligence*, NY: McGraw-Hill, 1982. (Based on Ph.D. thesis, Stanford University, CA, 1977.)

Lenat, D. "EURISKO: a Program that Learns New Heuristics and Design Concepts: the Nature of Heuristics, III: Program Design and Results." *Artificial Intelligence*, v. 21(2), 1983: 61–98.

Lewis, H. R. and C. H. Papadimitriou. *Elements of the Theory of Computation*. Englewood Cliffs, NJ: Prentice-Hall, 1981.

Lindsley, D. E., and J. E. Holmes. *Basic Human Neurophysiology*. NY: Elsevier, 1984.

Longo, G., ed. *Information Theory: New Trends and Open Problems*. NY: Springer-Verlag, 1975: iii–iv.

Lorenz, C. *The Design Dimension: Product Strategy and the Challenge of Global Marketing*. NY: Basil Blackwell, 1986.

Luce, R. D. "The Theory of Selective Information and Some of its Behavioral Implications." In R. D. Luce, ed., *Developments in Mathematical Psychology*, Glencoe, IL: Free Press, 1960: 5–119.

Luciano, D. S., J. H. Sherman, and A. J. Vander. *Human Anatomy and Physiology: Structure and Function*, 2nd ed. NY: McGraw-Hill, 1983.

Luria, S. E. *36 Lectures in Biology*. Cambridge, MA: MIT Press, 1975.

Machlup, F. and U. Mansfield, eds. *The Study of Information: Interdisciplinary Messages*. NY: Wiley, 1983.

Maley, J. G. "Managing the Flow of Intelligent Parts." *Robotics and Computer-Integrated Manufacturing*, v.4 (3/4), 1988: 525–530.

Mallet, F. L. "Microcomputer-Based Control of Composites Curing." S.M. thesis, MIT, Cambridge, MA: June, 1985.

Mandelbrot, B. *The Fractal Geometry of Nature*. NY: Freeman, 1977.

Manheim, M. L. *Hierarchical Structure: A Model of Design and Planning Process*. Cambridge, MA: MIT Press, 1966.

March, J. and H. Simon. *Organizations*. NY: Wiley, 1958.

Marten, D. W., Jr., P. A. Mayes, V. W. Rodwell, and D. K. Granner. *Harper's Review of Biochemistry*, 20th ed. Los Altos, CA: Lange Medical Publications, 1985.

McClelland, J. L., D. E. Rumelhart, and the PDP Research Group. *Parallel Distributed Processing, v.2: Psychological and Biological Models*. Cambridge, MA: MIT Press, 1986.

McCree, J. "Vented Compression Molding: A New Method for Molding Particle Materials." S.M. thesis, M.I.T., Cambridge, MA, Jan. 1983.

McDonald, Anthony C. *Robot Technology*. Englewood Cliffs, NJ: Prentice-Hall, 1986.

McEliece, R. J. *The Theory of Information and Coding*. Reading, MA: Addison-Wesley, 1977.

McLean, C., M. Mitchell and E. Barkmeyer. "A Computer Architecture for Small-Batch Manufacturing. *Spectrum*, v. 20(5), May, 1983: 59–64.

Mesarovic, M. D. and Y. Takahara. *Theory of Hierarchical, Multilevel Systems*. NY: Academic, 1970.

Mesarovic, M. D. and Y. Takahara. *General Systems Theory: Mathematical Foundations*. NY: Academic, 1975.

Michalski, R. S., J. G. Carbonell and T. M. Mitchell, eds. *Machine Learning: An Artificial Intelligence Approach*. Palo Alto, CA: Tioga Publishing Co., 1983.

Miller, J. G. *Living Systems*. NY: McGraw-Hill, 1978.

Minsky, M.L. *Computation: Finite and Infinite Machines*. Englewood Cliffs, NJ: Prentice-Hall, 1967.

Minsky, Marvin. *The Society of Mind*. NY: Simon and Schuster, 1986.

Mishkoff, H. C. *Understanding Artificial Intelligence*. Indianapolis, IN: Howard W. Sams and Co., 1985.

Mitchell, T., L. Steinberg, G. Reid, P. Schooley, H. Jacobs, and V. Kelly. "Representations for Reasoning About Digital Circuits." *Proc. International Joint Conference on Artificial Intelligence*, Vancouver, B.C, 1981.

Mizoguchi, F. "Prolog-based Expert System. "*New Generation Computing,* v. 1, 1983: 99–104.

Moravec, H. P. *Robot Rover Visual Navigation*. Ann Arbor, MI: UMI Research Press, 1981.

Moravec, H. P. *Mind Children: The Future of Robot and Human Intelligence*. Cambridge, MA: Harvard Univ. Press, 1988.

Mosher, R. S. "Exploring the Potential of a Quadruped." SAE paper 690191, International Automotive Engineering Conference, Detroit, MI, 1969.

Mostow, J. "Program Transformations for VLSI. " *Proc. International Joint Conference on Artificial Intelligence*, Karlsruhe, W. Germany, 1983.

Mostow, J. "Toward Better Models of the Design Process." *AI Magazine*, v. 6, Spring 1985: 44–57.

Mountcastle, V. B. "The Problem of Sensing and the Neural Coding of Sensory Events." In

*The Neurosciences: A Study Program*. Edited by G. C. Quarton, T. Melnechuk, and F. O. Schmitt. NY: Rockefeller Univ. Press, 1967: 393–408.

Nakazawa, H. and N. P. Suh. "Process Planning Based on Information Concept." *Robotics and Computer-Integrated Manufacturing*, v. 1(1), 1984: 115–123.

Newell, A. "The Knowledge Level." *Artificial Intelligenc Magazine*, v.18(1), 1982: 87–127.

Newell, A. and H. A. Simon. *Human Problem Solving*. Englewood Cliffs, NJ: Prentice-Hall, 1972.

Nilsson, N. J. "A Mobile Automation: An application of Artificial Intelligence Techniques." *Proc. First International Joint Conf. Artificial Intelligence*, Washington, D.C., May 1969, pp. 509–515.

Nilsson, N. J., and C. A. Rosen. "Application of Intelligent Automata to Reconnaissance." Technical Report, SRI, Menlo Park, CA, February 1969.

Noback, C. R., and R. J. Demarest. *The Human Nervous System*, 3rd ed. NY: McGraw-Hill, 1981.

Noback, C. R., and R. J. Demarest. *The Nervous System: Introduction and Review*, 3rd ed. NY: McGraw-Hill, 1986.

Oliveira, E. "Developing Expert Systems Builders in Logic Programming." *New Generation Computing*, v. 2(2), 1984: 187–194.

Padulo, L. and M. A. Arbib. *Systems Theory: A Unified State Space Approach to Continuous and Discrete Systems*. Philadelphia: Saunders, 1974.

Pao, Y.-H. "A Connectionist-net Approach to Autonomous Machine Learning of Effective Process Control Strategies." *Robotics and Computer-integrated Manufacturing*, v.4 (3/4), 1988, forthcoming.

Pao, Y.-H. and M. Jelinek. "Flexible Manufacturing Cells and Systems." In R. C. Dorf, ed., *International Encyclopedia of Robotics*, New York: Wiley, 1988:530–551.

Parthasarathy, S., and S. H. Kim. "Formal Models of Manufacturing Systems." Technical Report, Laboratory for Manufacturing and Productivity, MIT, Cambridge, MA: 1987.

Pascale, Richard T. and A. G. Athos. *The Art of Japanese Management*. NY: Warner, 1981.

Pereira, F., ed. "C-Prolog User's Manual," version 1.5. Dept. of Architecture, University of Edinburgh, UK, Feb. 1984.

Peters, T. *Thriving on Chaos*. NY: Knopf, 1987.

Poggio, T., and C. Koch. "Synapses that Compute Motion," *Scientific American*, v. 256 (5): 46–52.

Prigogine, I. *From Being to Becoming: Time and Complexity in the Physical Sciences*. NY: Freeman, 1980.

Prior, A. N. *Time and Modality*. Oxford: Clarendon Press, 1957.

Quastler, H., ed. *Information Theory in Psychology: Problems and Methods*. Glencoe, IL: Free Press, 1955.

Quillian, R. "Semantic Memory." In M. Minsky, ed., *Semantic Information Processing*, Cambridge, MA: MIT Press, 1968.

Raibert, Marc H., H. Benjamin Brown, Jr., Michael Cheppoonis. "Experiments in Balance with a 3D One-Legged Hopping Machine." *International Journal of Robotics Research*, v. 3(2), 1984b, pp. 75–92.

Raibert, Marc H. "Hopping in Legged Systems–Modeling and Simulation for the Two-Dimensional One-Legged case." *IEEE Trans. Systems, Man, Cybernetics*, v. 14(3), 1984a, pp. 451–463.

Raibert, Marc H. *Legged Robots that Balance*. Cambridge, MA: MIT Press, 1986.

Rescher, N. and A. Urquhart. *Temporal Logic*. NY: Springer-Verlag, 1971.

Rich, C., H. E. Shrobe, and R. C. Waters. "Overview of the Programmer's Apprentice." *Proc. Internation Joint Conference on Artificial Intelligence*, Tokyo, Japan, 1979.

Rich, E. *Artificial Intelligence*. NY: McGraw-Hill, 1983.

Rinderle, J. R. and N. P. Suh. "Measures of Functional Coupling in Design." *ASME Paper 82-PROD-27*, 1982.

Roberts, F. S. *Measure Theory, with Applications to Decisionmaking, Utility, and the Social Sciences*. Reading, MA: Addison-Wesley, 1979.

Robinson, J. A. "Machine Oriented Logic Based on the Resolution Principle." *J. ACM*, v. 12, 1965: 23–44.

Rowe, P. G. *Design Thinking*. Cambridge, MA: MIT Press, 1987.

Rumelhart, D. E., G. E. Hinton and J. L. McClelland. "A General Framework for Parallel Distributed Processing." In D.E. Rumelhart, J.L. McClelland and the PDP Research Group, *Parallel Distributed Processing, v.1: Foundations*, Cambridge, MA: MIT Press, 1986: 45–76.

Russell, B. *An Outline of Philosophy*. London: George Allen and Unwin, 1927: 292.

Russell, E. S. *Form and Function*. Chicago: The University of Chicago Press, 1916; rpt. 1982.

Sain, M. K. *Introduction of Algebraic Systems Theory*. NY: Academic, 1981.

Schank, R. C. *Dynamic Memory: A Theory of Reminding and Learning in Computers and People*. Cambridge, UK: Cambridge Univ. Press, 1982.

Scherlis, W. and D. Scott. "First Steps Toward Inferential Programming." *IFIP Congress '83*, North Holland, 1983.

Schmidt, R. F. "The Physiology of Small Groups of Neurons: Reflexes." In *Fundamentals of Neurophysiology*. Edited by R. F. Schmidt. Berlin: Springer-Verlag, 1985: 103–25.

Schrodinger, E. *What Is Life? and Mind and Matter*. Cambridge: Cambridge Univ. Press, 1967.

Sculley, J. with J. A. Byrne. *Odyssey: Pepsi to Apple*. NY: Harper and Row, 1987.

Shannon, C. E. and W. Weaver. *The Mathematical Theory of Communication*. Urbana: University of Illinois, 1949.

Shepherd, G. M. *Neurobiology*. NY: Oxford Univ. Press, 1983.

Shibazaki, H. and S. H. Kim. "Learning Systems for Manufacturing Automation: Integrating Explicit and Implicit Knowledge." Technical Report, Laboratory for Manufacturing and Productivity, MIT, Cambridge, MA: 1989.

Shortliffe, E. H. *Computer Based Medical Consultations: MYCIN*. NY: American Elsevier, 1976.

Simon, H. A. *Administrative Behavior*. NY: Macmillan, 1947.

Simon, H. A. *The Sciences of the Artificial*, 2nd ed. Cambridge, MA: MIT Press, 1985.

Smith, A. *An Inquiry into the Nature and Causes of the Wealth of Nations*, vols. 1,2, and 3, 10th edition. London: Caddell and Davies, 1802.

Smith, B. S. and M. Sciaky. "Robots in Welding." In R. C. Dorf and S. Y. Nof, eds., *International Encyclopedia of Robotics*, v. 3. NY: Wiley, 1988, pp. 1964–1978.

Spur, G. "Growth, Crisis and Future of the Factory." *Robotics and Computer-Integrated Manufacturing*, v. 1(1), 1984: 21–38.

Stefik, M. "Planning with Constraints (MOLGEN: Part 1 and Part 2)." *Artificial Intelligence*, v. 16(2), 1980.

Sterling, L. and E. Shapiro. *The Art of PROLOG*. Cambridge, MA: MIT Press, 1986.

Stone, J. and B. Dreher. "Parallel Processing of Information in the Visual Pathways: A General Principle of Sensory Coding." *Trends in Neuroscience*, v. 5, 1982: 441–47.

Suh, N. P. "The Future of the Factory." *Robotics and Computer-Integrated Manufacturing*, v. 1(1), 1984a: 39–49.

Suh, N. P. "Development of the Science Base for the Manufacturing Field through the Axiomatic Approach." *Robotics and Computer-Integrated Manufacturing*, v. 1 (3/4), 1984b: 397–415.

Suh, N. P. *The Principles of Design*. NY: Oxford Univ. Press, 1990.

Suh, N. P., A. C. Bell and D. C. Gossard. "On an Axiomatic Approach to Manufacturing and Manufacturing Systems." *Journal of Engineering for Industry*, Paris, France: Trans. ASME, v. 100(2), May, 1978a: 127–130.

Suh, N. P., S. H. Kim, A. C. Bell, D. R. Wilson, N. H. Cook, and N. Lapidot. "Optimization of Manufacturing Systems through Axiomatics." *Annals of the CIRP*, v. 27(1), 1978b: 383–388.

Summers, C. E. *Strategic Behavior in Business and Government*. Boston: Little, Brown, 1980.

Swift, K. G. *Knowledge-Based Design for Manufacture*. Englewood Cliffs, NJ: Prentice-Hall, 1987.

Szolovitz, P. "Artificial Intelligence and Clinical Problem Solving." Lab. for Computer Science, MIT, LCS-TM-140, Sept. 1979.

Tang, A. M., F. M. Westfield and J. S. Worley, eds. *Evolution, Welfare, and Time in Economics*. Lexington, MA: Lexington Books, 1976.

Therman, E. *Human Chromosomes: Structure, Behavior, Effects*, 2nd ed. Berlin: Springer-Verlag, 1986.

Thompson, D'Arcy. *On Growth and Form*, abridged edition. Edited by J. T. Bonner. Cambridge: Cambridge Univ. Press, 1984.

Thompson, J. M. T. and H. B. Stewart. *Nonlinear Dynamics and Chaos*. NY: Wiley, 1986.

Tokawa, T. "A Knowledge-Based System for Factory Configuration." S.M. Thesis, Massachusetts Institute of Technology, Cambridge, MA: May, 1989.

Toynbee, A. J. *A Study of History*. Abridgement of vols. I–VI by D.C. Somerville. London: Humphrey Milford and Oxford Univ. Press, 1947.

Tsypkin, Y. Z. *Adaptation and Learning in Automatic Systems,* trans. by Z. J. Nikolic. NY: Academic, 1971.

Tzu, Sun. *The Art of War*, trans. by S. B. Griffith. Oxford: Oxford Univ. Press, 1963. Also published as *The Art of Strategy*, trans. by R. L. Wing. NY: Doubleday, 1988.

Ulrich, K. T. and W. P. Seering. "A Computational Approach to Engineering Design Generation." *Robotics and Computer-Integrated Manufacturing*, v. 4, 1988, forthcoming.

Van Caneghem, M. and D. H. D. Warren. *Logic Programming and Its Applications*. Norwood, NJ: Ablex Publishing Corporation, 1986.

van Melle, W., A. C. Scott, J. S. Bennet and M. A. Peairs. "The EMYCIN Manual." Heuristic Programming Project, HPP-81-16, Stanford University, Nov. 1981.

Vander, A. J., J. H. Sherman, and D. S. Luciano. *Human Physiology: The Mechanisms of Body Function*, 4th ed. NY: McGraw-Hill, 1985.

von Bertalanffy, L. *General Systems Theory*. NY: Braziller, 1968.

von Bertalanffy, L. "The Theory of Open Systems in Physics and Biology." *Science,* v.111, 1950: 23–29.

von Mises, L. *Human Action: A Treatise on Economics*. New Haven: Yale Univ. Press, 1949.

von Mises, L. *The Ultimate Foundation of Economic Science*. NY: Van Nostrand, 1962.

von Neumann, J. *Theory of Self-Reproducing Automata*, ed. by A. W. Burks. Urbana: University of Illinois Press, 1966.

Walker, A. "Data Bases, Expert Systems, and Prolog." Report RJ3870 (44067), IBM Research Lab., San Jose, CA, 22 April 1983.

Warfield, J. N. *Structuring Complex Sysems.* Battelle Memorial Institute, Monograph No.4, Columbus, OH, April 1974.

Weaver, W. "Science and Complexity." *American Scientist,* v.36(4), Oct. 1948: 536–544.

Whitehead, A. N. *Modes of Thought.* NY: Macmillan, 1938.

Whyte, L. L., A. G. Wilson, and D. Wilson. *Hierarchical Structures.* NY: American Elsevier, 1969.

Wiener, Norbert. *The Human Use of Human Beings.* NY: Avon, 1967.

Wilson, D. R. "An Exploratory Study of Complexity in Axiomatic Design." PhD. thesis, MIT, Cambridge, MA, Feb. 1980.

Winchester, A. M., and T. R. Mertens. *Human Genetics*, 4th ed. Columbus, OH: Charles E. Merril, 1983.

Winograd, T. and F. Flores. *Understanding Computers and Cognition: A New Foundation for Design.* Norwood, NJ: Ablex, 1986.

Winston, P. H. and B. K. P. Horn. *LISP.* Reading, MA: Addison-Wesley, 1981.

Wright, P. F. and D. A. Bourne. *Manufacturing Intelligence.* Reading, MA: Addison-Wesley, 1988.

Yasuhara, M. "Axiomatic Engineering Design: Its Concept and Procedure." S.M. thesis, MIT, Cambridge, MA, Feb. 1980.

Yoshikawa, H. "Multi-purpose Modeling of Mechanical Systems—Morphological Model as a Mesomodel." In O. Bjorke and O. I. Franksen, eds., *System Structure in Engineering: Economic Design and Production,* Trondheim, Norway: Tapir, 1978: 594–629.

Yu, F. T. S. *Optics and Information Theory.* NY: Wiley, 1976.

# Name Index

# Subject Index

*Numbers in italics refer to figures or tables*